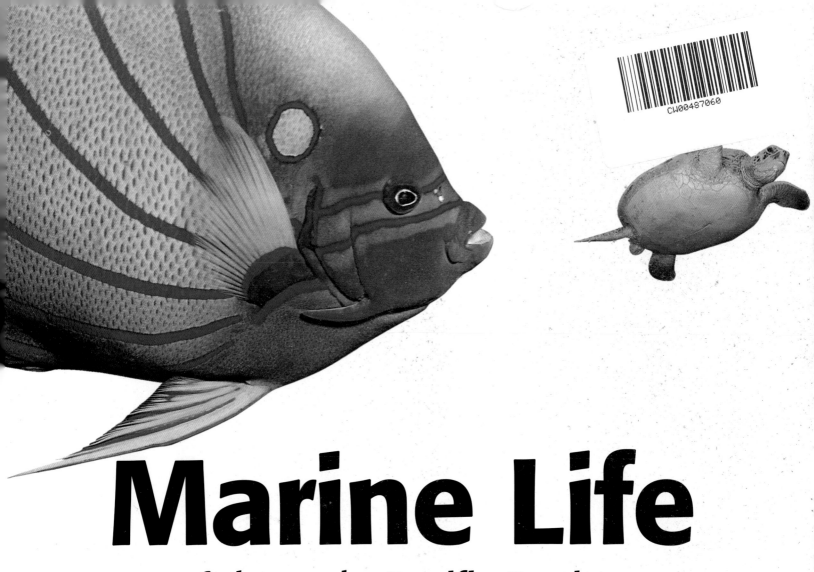

Marine Life

of the Indo-Pacific Region

including Indonesia, Malaysia, Thailand and all of Southeast Asia

Text by Gerald Allen, PhD
Photographs by Gerald Allen, Roger Steene, Rudie Kuiter, Mark Strickland,
Burt Jones/Maurine Shimlock and others
Edited by Fiona Nichols and Michael Stachels

PERIPLUS
EDITIONS

Publisher **Eric M. Oey**
Editors **Fiona Nichols and Michael Stachels**
Layout Design **Peter Ivey**
Cover Design **Momentum Design Pte Ltd**
Production **Mary Chia**

ISBN: 962-593-016-7
Printed in the Republic of Singapore

Address all inquiries and comments to
Periplus (Singapore) Pte Ltd
5 Little Road
#08-01
Singapore 536983

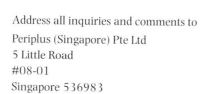

Acknowledgements
The author and editors would like to thank the
following people who provided assistance and
information during the production of this book:
Ron Holland, Graham and **Donna Taylor** of
Borneo Divers, **Dr. Hanny** and **Inneka Batuna** of
Manado Murex Resort, **Anton Saksono** of Pulau
Putri Resort, **Henrik Nimb** of PADI, Singapore,
and **Michael Lee** of Tioman Berjaya Resort.

contents

features

Text by Gerald Allen, PhD
Photographs by Gerald Allen,
Roger Steene, Rudie Kuiter,
Mark Strickland, Burt Jones,
Maurine Shimlock and others

Introduction

Site of The World's Greatest Biological Diversity

Welcome to the incredible world of Southeast Asian coral reefs. No other region on earth presents such a dazzling wealth of undersea life. If you have dived in places such as the Caribbean, Red Sea and Hawaii, but still crave the ultimate experience, look no further than Southeast Asia. Its extraordinary wealth of marine creatures is absolutely astounding. There are approximately three times as many reef-dwelling fishes and corals in the region compared with those of the Caribbean! A comparison of most other groups of plants and animals reveals similar figures.

Why is there such a wealth of marine life in Southeast Asian seas compared with other parts of the globe? Several important factors have played a role. The combination of a huge, shallow continental shelf constantly bathed by warm seas has provided a fertile stage on which evolution has proceeded. Evolution is a very complex process but in simplest terms consists of the interplay of genetic and environmental forces on a population of organisms, be it a fish, worm or human being. Over time, usually measured in thousands of years, the organism is moulded or adapted to survive more efficiently in its environment. Failure to adapt ultimately leads to extinction.

During the course of evolution a population may be split up and isolated for various reasons. If the isolation persists each population will gradually develop into a separate species. A multitude of isolating mechanisms have no doubt contributed to the diversity of species in Southeast Asian seas. Periodic raising and lowering of sea levels, particularly over the past 25,000 years, have created land barriers, cutting off adjacent marine populations. The island nature of the region has also created numerous opportunities for isolation of shallow water organisms. Indonesia alone has over 15,000 islands. The discharge of numerous silt-laden rivers also forms barriers that may divide populations of coral reef plants and animals. The discharge patterns show dramatic changes over time in response to fluctuation in sea levels and therefore the effects may be very complex and interesting.

Last, but not least, the tremendous variety of shallow marine habitats has fostered the evolution of a matching number of specialised plant and animal communities. Although coral reefs are by far the richest of these environments, nearby mangrove, seagrass, seaweed, mud, sand and rubble habitats all have characteristic organisms that often interplay with those of the reef itself.

Coral reefs are indeed nature's richest realm. They are extremely complex systems consisting of numerous micro habitats. The huge number of species found on coral reefs is a direct reflection of the high number of habitat opportunities afforded by this environment. In addition to the obvious community of plants and ani-

mals seen on or above the reef's surface, there are thousands of unseen organisms. Diverse communities live under rocks and dead coral slabs, or in the crevices and fissures of the reef. An incredible number of species are associated with live and dead coral heads. A single head may contain more than 100 species of worms and a numerous assortment of other organisms!

Symbiotic associations, in which a pair of unrelated organisms live together for the benefit of either or both partners' benefit, are also extremely common. For example, sponges, soft and hard corals, echinoderms, and sea squirts frequently have crustaceans, molluscs, worms and fishes living on their outer surface or within internal cavities. In addition, a community of macroscopic and microscopic animals live below the surface of sand and rubble bottoms. Finally, there is a legion of microscopic creatures that live either on the reef's sur-

Opposite and below: Southeast Asia has an extraordinary wealth of coral reefs and beautiful island scenery.

ROGER STEENE

Types of Coral Reefs

Coral reefs are an incredibly diverse phenomenom. No two reefs are exactly the same. Each has its own peculiarites with regards to such features as bottom topography, water clarity, current patterns and life forms. At first glance the variety is amazing, seemingly endless, but reefs can be broadly classified into one of three main types depending on their origin: atolls, fringing reefs and barrier reefs.

All three main types of reefs are found in Southeast Asia. Of the three, fringing reefs are by far the most common. They occur along the shoreline of continents and around islands. They are built on the shallow submerged margin of the landmass, continuing to grow upward and outward in tune with either rising sea level, sinking land mass, or combination of the two. Barrier reefs are similar in structure but separated from the island or mainland by an extensive lagoon.

There are few examples of well-developed barrier reefs in the region (Australia's Great Barrier Reef is the world's best example) as most are fragmented and relatively small.

Atolls are also scarce, the largest in Southeast Asia, Taka Bone Rate, is situated between Flores and Sulawesi. It was Charles Darwin who was the first to offer a logical explanation of how atolls are formed, following his 1836 visit to the Cocos Keeling Islands. An atoll is actually the last part of a fringing reef that once surrounded the shallow margin of a volcanic island. As the island gradually sinks (a few millimetres annually) the living corals grow upward in successive layers to maintain contact with optimum light conditions. Eventually the island disappears in the sea, but the fringing reef persists as a ring around a central lagoon.

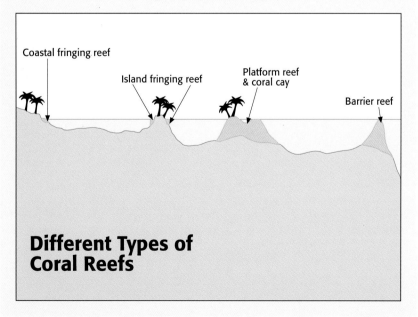

Coastal fringing reef

Island fringing reef

Platform reef & coral cay

Barrier reef

Different Types of Coral Reefs

face or in mid-water directly above. The latter are referred to as zooplankton, a mixture of larval and adult forms, that are an important food source for many reef inhabitants.

The Basics...Clear Water and Sunlight

Coral reefs are composite structures created from a wide variety of living and dead plants and animals. However, they owe their existence mainly to the reef building hard corals. Individual coral animals or polyps secrete a calcium carbonate skeleton that forms the matrix of the reef. Reef-building varieties of coral can only survive in warm, relatively clear sea water. The richest growths are found in equatorial seas. In general, they do not grow at latitudes where the temperature is greater than 30°C or where the minimum winter sea temperature drops below 20°C.

Corals thrive with maximum exposure to sunlight, therefore clear water is an important requirement. Ironically, the clear blue seas typical of the tropics are very poor in nutrients. They are the marine equivalents of deserts. Fortunately, thanks to algae growing both on the surface of the reef and within the cells of coral and other invertebrates, coral reefs are largely self-sustaining and are very efficient at recycling the nutrients they produce.

Structure of Coral Reefs

Coral reefs result from a combination of constructive and destructive processes. The beautiful shapes and colours of corals that we generally think of as the reef itself actually represent a small portion of its overall structure. The living coral forms a thin veneer over a platform composed of the remains of previous coral generations. In some cases the platform may be hundreds of metres thick. Corals respond to sunlight by growing upwards.

Modern reefs have developed mainly over the past 5,000 years in response to rising sea levels which in turn were the result of melting glacial ice left from the last great Ice Age. The rate of this rise has been approximately 1-2 metres per 1,000 years. It has been estimated that an upward coral growth of one metre can take anywhere between 300-3,000 years. In addition to their upward growth, reefs also grow in an outward or seaward direction at similar rates.

Construction of the reef platform depends on the interplay of a host of destructive forces—both natural and animal. Worms and other boring organisms attack living corals. They continue their activities even after the corals are dead. Cyclonic storms and earthquakes also help to fragment the coral skeletons. These particles combine with the glass-like, calcareous spicules or splinters, and shells of other dead animals to form the basic limestone building blocks for successive layers of the platform. Green *Halimeda* algae has a high content of calcium carbonate and its dead remains also contribute significantly to the bottom sediments. All these materials are eventually cemented into a solid limestone structure by encrusting red coralline algae and magnesium calcite particles that are precipitated from sea water.

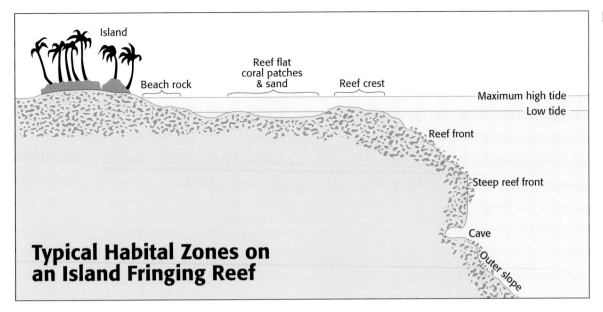

Typical Habital Zones on an Island Fringing Reef

Island

Beach rock

Reef flat coral patches & sand

Reef crest

Maximum high tide
Low tide

Reef front

Steep reef front

Cave

Outer slope

Cross Section of a Typical Coral Reef

This is a simplified diagram of a typical fringing reef showing the location of the major habitat zones. Corals and fishes are most abundant on the reef front, the slope of which is extremely variable. Corals are generally more sparse and assume a stunted, thickset growth form on the wave-swept reef crest. Reef flats range in width from a few metres to several hundred metres.

Plant and Animal Classification Made Simple

There's no need to panic at the sight of the unpronounceable scientific names that appear in this book. Tackle them one syllable at a time. And don't feel self conscious about mispronouncing them. There is seldom agreement on their pronunciation, even among scientists! They are part of a clever classification system introduced by the Swedish botanist, Carole Linnaeus, in the mid-18th century. At the heart of this system is the binomial or double word that pertains to the genus and species. Every described organism, be it a seaweed, crab, fish or human being has a two-part scientific name that is usually composed from Latin or Greek words. The first part is the genus or generic name and the second is the species or specific name (these names are generally italicised).

According to its relationships, based on similarities of both external and internal anatomy, an organism is assigned to the major categories or rungs on the overall ladder of classification.

For example, similar genera (plural of genus) are placed in the same family. Family names are given for most of the plants and animals in the identification section of this book. They can be recognised by their ending which consists of -dae in animals and -aceae in plants. Related families are placed in the same order. The highest rungs on the classification ladder pertain to Class and Phylum.

The classification of the Spine-cheek anemonefish (see page 79) is presented as an example:

PHYLUM: Chordata (all animals with a notochord)
CLASS: Osteichthyes (all bony fishes)
ORDER: Perciformes (most reef fish families)
FAMILY: Pomacentridae (all damselfishes)
GENUS: *Premnas* (Spine-cheek anemonefish, only species in this group)
SPECIES: *biaculeatus* (Spine-cheek anemonefish)
As you will see in using this (or any other scientific) book, animals or fish are usually referred to either by their English common names or by the two Latin names denoting genus and species.

Below: Spine-cheek anemonefish (*Premnas biaculeatus*).

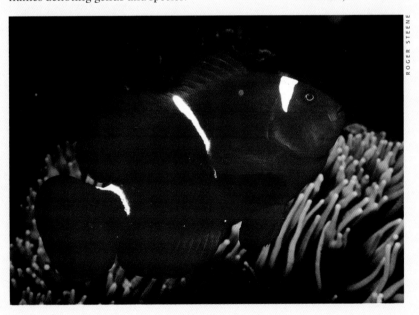

ROGER STEENE

Marine Plants
Fundamental to the Reef

Marine plants are frequently ignored in popular guidebooks but deserve at least token coverage. It would be virtually impossible for animal life to exist without them. Like their terrestrial counterparts, marine plants utilise sunlight and carbon dioxide to photosynthesise energy-rich compounds: carbohydrates and sugars. These provide direct nutrition to a wealth of plant-eating microorganisms as well as larger animals, particularly molluscs, crustaceans and fishes. Even though the purely carnivorous species never consume plants, they owe their existence to complex predator-prey interactions involving plants, herbivores and smaller carnivores. Plants are therefore the fundamental unit of the coral reef "food web".

Plants also provide shelter for diverse animal communities. Fishes, usually camouflaged, utilise plant cover either to hide from predators or to sneak up on prey. In addition, a host of invertebrates live on the surface of seaweeds and other marine plants. A few nudibranchs have carried this association to the extreme. They feed on certain green algae incorporating the plant's pigment cells or chloroplasts into their own tissue, which impart an effective camouflage.

Two major types of plants are encountered in coral reef environments in Southeast Asia. Seaweeds or algae, although not nearly as dominant in the tropics as in temperate latitudes, are well represented. They are quite different compared to terrestrial plants, lacking true leaves, stems and roots, although they frequently have parts that resemble these structures. The second major category is the seagrasses, which are the only type of vascular plant found in the open sea. They possess both flowers and fruits, and also set seeds.

Seaweeds

The main types of seaweeds can usually be identified on the basis of colour but there are some confusing exceptions. Scientists use this same scheme in classifying seaweeds into three categories: green, brown and red. All contain the green pigment chlorophyll, but brown and red algae have additional pigments that either mask or blend with the green ones. All three types are extremely variable with regards to size and growth form. Some are globular and sponge-like, many others are branching and bushy, still others form slender filaments. There are also numerous species, such as the red encrusting coralline algae and green *Halimeda* algae, that incorporate calcium carbonate into their internal framework, giving them a brittle texture.

Seagrasses and Mangroves

Although they really form separate ecosystems, small pockets of mangrove or seagrass are often present on coral reefs, particularly if situated close to shore. They form important refuge or nursery areas for many fishes and invertebrates. For example, several snappers and emperors depend on seagrass beds for food and protection during their vulnerable youthful stages.

Mangrove systems can be extremely productive because they accumulate or trap nutrients from both land and sea. Several species of commercially important fish and prawns breed and spend their critical first stages of life in mangrove estuaries. Unfortunately, the mangrove environment is increasingly threatened by coastal land reclamation and silting due to logging.

Forams

Forams are unrelated to plants and form a distinct group or phylum, the Foraminiferida. They are single-celled animals consisting of a jelly-like blob of protoplasm surrounded by a calcium carbonate shell. There is also a thin outer layer of protoplasm with projections or pseudopods used for locomotion, feeding, waste expulsion and gas exchange. Most types are very small, usually less than 1–2 cm in diameter.

Reef species are either free living on the bottom or are attached to other organisms, particularly seagrasses and seaweeds. So abundant are these tiny animals that it is estimated their shells form 50 percent of the ocean's calcareous bottom sediments. The shells come in a huge variety of shapes. The beach at Padang Bai, on the Indonesian island of Bali, is composed of nearly 100 percent forams.

1 Sea Grapes (5 cm height)
Caulerpa racemosa; Chlorophyta (Green Algae)
Several species of *Caulerpa* green algae are present on coral reefs. Although a variety of growth forms are evident, they can be recognised by their characteristic interconnecting system of creeping roots or rhizomes. The species shown here is readily identified by its grape-like vesicles.

2 Green Coralline Alga (clump width 15 cm)
Halimeda capiosa; Chlorophyta (Green Algae)
Halimeda algae are frequently the most conspicuous plants on coral reefs. Individual plants consist of branching chains of flattened or cylindrical, leaf-like brittle segments. The white, wafer-thin remains of dead *Halimeda* are a conspicuous part of bottom sediments that surround many reefs. The species shown here is found on sand bottoms in contrast to most other *Halimeda*, which usually grow on hard surfaces.

3 Sailor's Eyeball (6 cm)
Ventricaria ventricosa; Chlorophyta (Green Algae)
These curious structures invariably prompt the question, "What is it?" from divers and snorkellers. It is actually a very uncomplicated type of green alga which is largely composed of a single, very large fluid-filled cell

that has a rigid outer covering. Eyeballs are very common on shallow, protected reefs.

4 Turtle Weed (40 cm)
Chlorodesmis fastigata; Chlorophyta (Green Algae)
A brilliant green alga, often attracting divers, it typically grows in small clumps (20-50 cm width) on shallow reefs and is supposedly eaten by turtles.

5 Red Calcareous Alga (50 cm width)
Neogoniolithon brassica; Rhodophyta (Red Algae)
Various types of encrusting red calcareous algae are common on the surface of reefs, particularly on the upper part of outer slopes and on the reef crest. They often form a pink or red coloured coating on rocky bottoms or dead coral. These types of algae play an important role in the reef building process by binding loose bottom sediments.

6 Sargassum Weed (to 1.5 metre height)
Sargassum sp.; Phaeophyta (Brown Algae)
Thick growths of brown Sargassum algae may occur in close proximity to coral reefs. Bits and pieces constantly become detached as a result of wave action, resulting in the formation of floating rafts which can be huge and may drift well offshore. If you find a raft, it is well worth exploring as they usually teem with life. Crabs, shrimps, tiny molluscs and juvenile fishes are particularly abundant.

7 Broadblade Seagrass (20 cm height)
Enhalys acoroides; Potamogetonaceae (Seagrasses)
Small pockets of seagrass are sometimes seen on coral reefs, but the best examples are found away from this environment. Vast meadows exist in many coastal areas of Southeast Asia. Seagrass meadows are extremely productive, usually more so than comparable plots of agricultural land. Ironically, very few marine animals take advantage of this copious food supply; notably those that do are sea urchins, molluscs, turtles, dugongs and a few fishes.

8 Mangrove Tree (12 m maximum height)
Rhizophora stylosa; Rhizophoraceae
Situated near coral reefs, mangroves form a distinct ecosystem. Their dense foliage and roots provide shelter for scores of organisms. Numerous fishes and invertebrates spend at least part of their life cycle in this environment. This species has a cigar-shaped hypocotyl that grows from the attached fruit. When the fruit drops the sharp tip of the hypocotyl becomes firmly embedded in the mud and a new plant develops.

9 Disc Foram (each disc 1 cm)
Marginopora vertebralis; Miliolina
Forams are very common but seldom noticed. This species occurs on reef flats and in lagoon environments. The disc-shaped shell is well calcified and is often conspicuous among beach sediments.

1 Sea Grapes *Caulerpa racemosa*

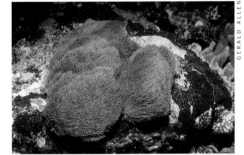
4 Turtle Weed *Chlorodesmis fastigata*

7 Broadblade Seagrass *Enhalys acoroides*

2 Green Coralline Alga *Halimeda capiosa*

5 Red Calcareous Alga *Neogoniolithon brassica*

8 Mangrove Tree *Rhizophora stylosa*

3 Sailor's Eyeball *Ventricaria ventricosa*

6 Sargassum Weed *Sargassum sp.*

9 Disc Foram *Marginopora vertebralis*

Sponges
One of Nature's Simplest and Most Ancient Organisms

It may be difficult to get excited about a lowly sponge, but this group has plenty to offer. Not only are they among the most colourful of the reef's creatures, but they harbour a wealth of interesting symbionts (different animals living together) and play an important role in medical research.

The sponges we see today on coral reefs also serve as a reminder of their glorious history. Sponge-like animals existed as far back as Precambrian times or approximately 600–700 million years ago. Some 400 million years ago sponges became the dominant form of undersea life and were important reef builders, a role now largely taken over by hard corals.

Sponges have the simplest structure of all multi-celled marine animals. The basic body plan consists of two layers of cells with a fibrous matrix sandwiched in between. The body is riddled with numerous holes or ostia through which water passes into a series of internal canals. These canals are partially lined with special collar cells or choanocytes. Each collar cell has a whip-like tail or cilia that projects into the water-filled canal. Vigorous movement of these tails creates a current that moves food-laden water through the sponge. This movement also creates a vacuum effect, drawing the surrounding water into the ostia. The water, along with the sponge's waste products, is eventually expelled through large conspicuous openings or oscula. A typi-cal sponge pumps water equal to at least four or five times its own volume every minute. For a football-sized sponge this translates into a quantity of several thousand litres each day!

The fibrous inner matrix is impregnated with numerous glass-like slivers of silica or calcium carbonate, which are called spicules. These form a loose skeletal structure, which in combination with the fibres, are responsible for the sponge's shape and rigidity. Sponge specialists rely on microscopic examination of the spicules to determine species identification. In many sponges the spicules are in the shape of needle-like rods with pointed ends. Other species have multi-pronged, star-like clusters of needles, and a few have an anchor-shaped arrangement.

Sponges are extremely efficient filter feeders. Water that is drawn in is subsequently filtered by a series of sieves of diminishing mesh size. They can filter the smallest of microscopic organisms and are particularly efficient in filtering bacteria, a major source of nutrition. Food particles are ingested by special mobile cells in the fibrous matrix. Because of their predilection for bacteria and other organic debris, sponges are often abundant and robust at the entrance of harbours or off river mouths. They are also prominent in areas exposed to vigorous currents which greatly facilitates the movement of water into the feeding canals.

Numerous sponges found on coral reefs are impregnated with the living tissue of blue-green algae. This symbiotic relationship is similar to that found between different types of algae and various corals, molluscs and ascidians. The algal cells carry on photosynthetic activities and leak a significant amount of energy-rich sugar compounds directly into the sponge tissue. In some types of sponge the algae may supply nearly 100 percent of the nutritional requirements.

An estimated 10,000 species of sponges are known. All but a few freshwater species are found in the sea. Perhaps as many as 2,000 species inhabit Southeast Asian reefs. Remarkably, only a small portion of these have been given official scientific names. The identification of tropical species is frequently difficult due to the extremely variable appearance that a single species of sponge assumes depending on environmental factors. For example, a particular species found in a protected lagoon may exhibit an entirely different shape and colour from the same species situated in an area of strong current on the outer reef.

Sponges have unusual biochemical properties and are frequently toxic to other organisms. They are therefore attractive subjects for medical research. Hundreds of species have already undergone biochemical analysis. It is hoped that substances will ultimately be found in these ancient organisms that will inhibit cancerous growths or other diseases.

Below: The giant Barrel sponge *(Xestospongia testudinaria)* is common throughout the region. It can grow to over a meter high.

ASHLEY BOYD

1 Solar Sponge (20 cm)
Mycale sp.; Mycalidae

The brilliant orange Solar sponge inhabits steep drop-offs exposed to strong currents in about 10–50 metres depth. It is frequently seen in caves or under ledges in association with *Dendronephthya* soft coral and sea squirts (*Didemnum*). Solitary individuals are common but it also occurs in clusters. It generally forms globular encrustations and shape is extremely variable.

2 Blood Sponge (50 cm)
Echinodictyum sp.; Raspailiidae

This sponge is usually seen on outer reef slopes with periodic strong currents, normally at depths of between 10–40 metres. Although somewhat variable in shape, it generally forms an open vase that is tapered at the base. Underwater it appears to be dark greenish-brown, but if illuminated or brought to the surface it is blood red.

3 Stalked Bumpy Sponge (30 cm)
Asteropus sarassinorum; (Coppatiidae)

This sponge has a very characteristic club shape consisting of a narrow stalk and large globular head. The surface is covered by numerous rounded protuberances. It grows on protected coastal reefs where there is heavy siltation, usually in depths less than 10 metres. The stalks are useful for keeping the sponge clear of the bottom and allowing for movement with the currents, preventing accumulation of debris on its outer surface.

4 Plate Sponge (35 cm)
Phyllospongia lamellosa; Spongiidae

This is a thin leathery sponge that has a relatively rough upper surface and smoother, lighter coloured underside. The depth range is generally from 3–15 metres, often where there is moderate wave action. A small gobiid fish (*Phyllogobius platycephalops*) usually shelters on the underside of the "leaves". Apparently it is rarely found without its symbiotic partner.

5 Stalagmite Sponge (15 cm)
Reniochalina stalagmitis; Axinellidae

This sponge is recognised by its brilliant orange colour and very rough texture. Growth forms are variable but branching columns are perhaps the most common. The habitat consists of semi-protected coastal reefs where there is moderate siltation at depths between 8–20 metres. It usually occurs as small, isolated clumps.

6 Veined Sponge (5 cm)
Crambe sp.; Esperiopsidae

Some of the encrusting sponges have incredibly intricate patterns. This species is often seen while diving in caves or shipwrecks where light levels are low. It appears as a grey-brown mat until illuminated with a torch. Numerous microscopic incurrent pores or ostia are situated in the red tissue. After the water is filtered, it passes into the interconnecting network of the white canals which empty into the larger excurrent openings.

1 Solar Sponge *Mycale sp.*

2 Blood Sponge *Echinodictyum sp.*

3 Stalked Bumpy Sponge *Asteropus sarassinorum*

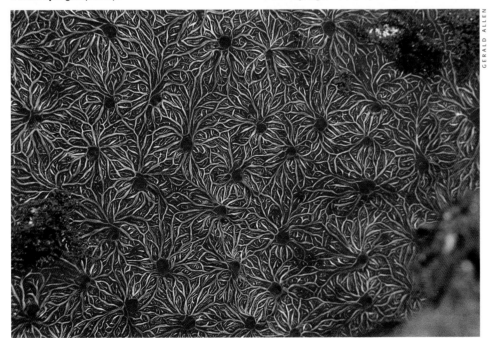

6 Veined Sponge *Crambe sp.*

4 Plate Sponge *Phyllospongia lamellosa*

5 Stalagmite Sponge *Reniochalina stalagmitis*

SPONGES

7 Pink Tube Sponge (40 cm)
Cribrochalina olemda; Niphatidae

This sponge occurs in lagoons or on protected reef slopes, in depths between 10 and 30 metres. Although somewhat variable in shape it generally forms clusters of elongate tubes that reach 20–40 cm in length. The relatively smooth-textured outer surface is occasionally covered with synaptid sea cucumbers. Colour ranges from lavender and pink to grey or bluish hues.

8 Barrel Sponge (70 cm)
Xestospongia testudinaria; Petrosiidae

This is perhaps the most readily recognised sponge on coral reefs due to its large size and characteristic barrel shape. The outer surface consists of narrow ridges and deep valleys. Specimens measuring one metre in height are common, but rare examples may reach twice this size. It is often covered with numerous small white sea cucumbers (*Synaptula*) as is shown in the photograph reproduced below.

9 Spiky Sponge (60 cm width)
Callyspongia sp.; Callyspongiidae

This species forms vase-like structures in both inshore and outer reef habitats. It is generally blue-grey in colour with a rough spiky outer surface. Sponges are frequently used by many other animals, especially echinoderms, crustaceans and gobiid fishes, for shelter and a source of accumulated debris consumed as food.

10 Curtain Sponge (70 cm)
Ianthella basta; Ianthellidae

The curtain sponge is variable in both colour and form. It may form vase-like growths or a rippled curtain that generally faces into the current. Yellow or purple varieties are commonly seen. The sponge lives in a variety of habitats but seems to thrive best on the floor of lagoons, or in passages where there is often turbid water and relatively heavy siltation. The surface is frequently covered with small white sea cucumbers of the genus *Synaptula*, which feed on accumulated organic debris.

11 Yellow Fan Sponge (50 cm)
Ianthella flabelliformis; Ianthellidae

Members of the family Ianthellidae or fan sponges, are unusual in lacking a skeleton composed of hard spicules but are supported by a network of thick fibres. This sponge is usually found in areas exposed to periodic strong currents. The ostia or incurrent pores generally face in the direction of the strongest prevailing currents.

12 Tennis Ball Sponge (8 cm)
Cinachyra sp.; Tetillidae

Tetillidae sponges generally have rounded growth forms and frequently exhibit bright, yellow-orange colours. This tennis ball-sized species has crater-like depressions on its surface. It is very common throughout the region at depths between 5–20 metres.

7 Pink Tube Sponge *Cribrochalina olemda*

10 Curtain Sponge *Ianthella basta*

11 Yellow Fan Sponge *Ianthella flabelliformis*

8 Barrel Sponge *Xestospongia testudinaria*

9 Spiky Sponge *Callyspongia sp.*

12 Tennis Ball Sponge *Cinachyra sp.*

Corals

The Reef Builders

Corals belong to the Phylum Cnidaria, a major division of the animal kingdom which also contains jellyfish, hydroids, anemones and sea fans. It seems a remarkably diverse group, but all its members share certain anatomical features. The basic body plan is characterised by radial symmetry. The body has three cell layers—an outer ectoderm and inner endoderm, with a primitive mesoderm, the jelly-like mesogloea, sandwiched in-between. Coelenterata, the former name for these animals, means "hollow gut", which describes the very basic internal anatomy. A single opening mostly serves the function of both mouth and anus.

The term "coral" is reserved for cnidarians that produce a hard skeleton composed of calcium carbonate or at least have a supporting structure of fibrous tissue that is reinforced with calcium or silica particles. The main types of corals include hydrocorals, hard corals (stony corals) and soft corals (octocorals). Hydrocorals have a hard skeleton, but are actually close relatives of hydrozoans (see p. 23). The other major types of corals—hard and soft corals—are described below.

Corals are capable of asexual reproduction by budding or fragmentation of the parent body. The existing polyps divide to form new ones, resulting in remarkably rapid growth. The fastest growing staghorn corals may add as much as 15 cm per year to their branching tips. They can also reproduce by sexual means, sometimes resulting in spectacular mass spawning events.

Anatomy of a Coral

Most corals, both soft and hard varieties, are colonial organisms composed of numerous individual polyps. The polyps are simple, fleshy structures nearly identical in composition to a sea anemone. One of the most remarkable aspects of their anatomy is the presence of microscopic, unicellular plants or algae that live within the tissue.

These algae, known as zooxanthellae, belong to a group called the dinoflagellates. They possess tiny projecting filaments that are vigorously moved to provide locomotion. This action probably is used to invade the tissue, but once embedded, there is little movement. The zooxanthellae thrive in this environment and the relationship is mutually beneficial. The algae utilise the coral's waste products, which combined with sunlight enables it to photosynthesise vital nutrients that are leaked into the surrounding tissues of the coral polyp. Although the polyps are well equipped with stinging cells to capture their own food, most reef-dwelling corals are absolutely dependent on the zooxanthellae to provide the bulk of nutritional requirements.

Zooxanthellae are not exclusive "room-mates" to hard and soft corals. They are impregnated in the tissues of a variety of other soft or fleshy invertebrates. Host animals include sponges, anemones, molluscs and sea squirts. The specialised algae are yellow-brown or greenish in colour and actually impart these same tones to the host animal's flesh. In general, most of the dull brownish or greenish corals host zooxanthellae, while brilliantly coloured types such as *Dendronephthya* lack them.

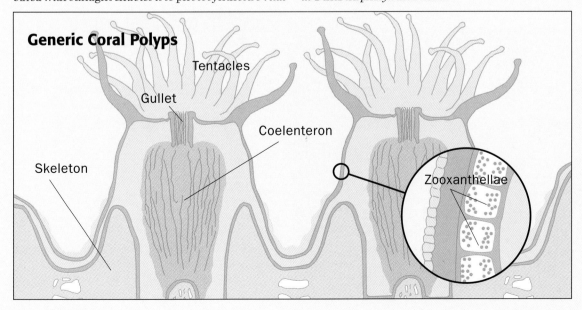

Generic Coral Polyps

Tentacles

Gullet

Coelenteron

Skeleton

Zooxanthellae

CORALS

Below (top): The unusual Hollow-branch gorgonian *(Solencaulon sp.)* has broad flattened blades with polyps on one side only.
Below (bottom): The tiny, flower-like polyps of this brilliant soft coral *(Echinogorgia)* are fully extended when feeding.
Opposite: The bright orange polyps of the Daisy coral *(Tubastraea faulkneri)* are usually retracted during the day. Photo by Jones/Shimlock.

Hard Corals (scleractinians)

Approximately 500 coral species are known from the Indo-Pacific region. The richest area is centred around Southeast Asia and neighbouring Australia where about 70 percent of the total species are found. Many corals have broad distribution patterns that extend the vast distance from the Red Sea and East Africa to islands of the Central Pacific. Although there are solitary types such as the mushroom corals (Fungiidae), the vast majority of species are colonial forms. In the latter category the members of the genus *Acropora*, frequently referred to as table corals and staghorn corals, are especially abundant.

An individual coral polyp consists of a fleshy sack topped with a ring of tentacles around a central mouth opening. It sits in a limestone skeletal case, which is actually secreted by the polyp. Members of the colony are linked by living tissue. Therefore nutrients captured by a section of the colony can be shared around. Many corals have brownish coloured, unicellular algae or zooxanthellae (see box opposite) living within the tissues of the polyps. They are called hermatypic corals. The algae use sunlight and carbon dioxide to produce carbon enriched organic compounds. These in turn are leaked to the polyp and may provide as much as 98 percent of its nutritional requirements.

Hard corals thrive in clear, warm shallow seas. Most depend on exposure to bright sunlight for optimal growth. This is why corals are most abundant and have their most spectacular growth forms in relatively shallow water (5–20 metres depth). Abundant growth may occur even shallower in sheltered locations provided there is minimal exposure to surge, for example in lagoons and protected inlets. The abundance of corals, both soft and hard, gradually diminishes with increasing depth below the optimum growth zone. Many of the deeper water species are "ahehermatypic" forms—species without symbiotic zooxanthellae living in their tissues. Nutritional needs are provided by capturing planktonic prey with their stinging cells.

Corals can reproduce by sexual means resulting in the production of eggs and sperm. Nature's ways are often clever. The chance mingling of gonadal products is absolutely maximized by spectacular mass spawning. Although this phenomenon is best documented on the northern Great Barrier Reef of Australia, it occurs on reefs throughout Southeast Asia. In northern Queensland, prodigious numbers of eggs are released for several consecutive nights after the full moon in November. Night dives are a memorable experience – millions upon millions of freshly liberated eggs float slowly to the surface accumulating there in a rich froth. Swimming among the rising eggs with an underwater light is like being in the midst of an upside-down snowstorm. Once fertilized, the eggs gradually develop into planula larvae, which live near the surface for a variable period —days or weeks—before settling onto the reef to begin a new colony—assuming they escape the hordes of plankton consuming organisms.

Soft Corals (alcyonarians)

Alcyonarians are structurally similar to hard corals. Both contain colonies of polyps that gather planktonic food. However, as their name suggests, soft corals lack a hard limestone skeleton. Instead, the supporting stem consists of fleshy tissue that is reinforced by a matrix of microscopic calcium or silica particles (sclerites). Other soft corals, including gorgonian fans and sea whips, have a central core or axial skeleton composed of a horny proteinaceous substance. The polyps are embedded in a cork-like outer bark. Soft corals are found in all reef habitats, but achieve their most impressive growth forms in deeper parts of the reef, between about 10–30 metres depth. Foremost in this respect is the gorgeous multi-hued *Dendronephthya*.

There is an amazing diversity of soft corals. Many of the familiar soft, fleshy forms occuring on shallow reefs of Southeast Asia belong to the family Alcyoniidae. Another major group, gorgonians or "sea fans", is actually composed of several families and contains the largest growth forms. Some species span 4–5 metres.

Soft corals seldom have encrusting growths of algae, sponges and sea squirts on their surface. Their numerous feeding polyps help prevent these "undesirables" from settling. They also have various chemical secretions that prevent or inhibit marine growths.

RUDIE KUITER

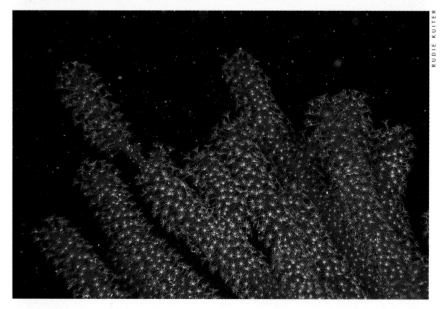

RUDIE KUITER

CORALS

1 Corallimorpharian (30 cm width)
Amplexidiscus fenestrafer; Actinodiscidae
These flattened disc-shaped animals are related to anemones. The upper surface is covered with a bed of tentacles. They can quickly envelop small fish prey, taking on a smooth globular shape in the process. Due to their similarity to host anemones it has been suggested they may lure and devour young anemonefishes that are searching for a home.

2 Zoantharian (colony width 8 cm)
Palythoa sp.; Zoanthidae
These colonial cnidarians are easily mistaken for soft corals but they are a distinct group, more closely related to hard corals. The individual polyps are interconnected by a tube-like structure. They do not secrete a calcareous skeleton like hard corals, but some species, such as the one shown here, incorporate bits of bottom sediment into their soft tissues during growth.

3 Staghorn Coral (300 cm width)
Acropora formosa; Acroporidae
Acropora is by far the most dominant coral on Indo-Pacific reefs. There are more than 100 species, of which at least 70–80 are found in our region. Several species have a characteristic branching growth form and are commonly known as staghorn corals. They flourish in relatively still waters of lagoons or deeper sections of fringing reefs.

4 Warty Coral (30 cm)
Pocillopora verrucosa; Pocilloporidae
The genus *Pocillopora* is easily recognised by the protruding wart-like growths or verrucae covering the surface. This species is commonly found in a variety of shallow water environments from fringing coastal reefs to wave-exposed outer reefs. It shows very pronounced changes in growth form depending on the amount of wave action. For instance, colonies living in areas of strong surge are relatively stunted and robust.

5 Creamy Coral (200 cm width)
Stylophora pistillata; Pocilloporidae
This species is common on reefs constantly buffeted by waves. It often forms extensive gardens on the upper edge of outer reef slopes. Colonies typically have thick, blunt branches with a relatively smooth texture. The colour is frequently creamy, but colonies are also grey, green, pink, or blue.

6 Montipora Coral (200 cm width)
Montipora sp.; Acroporidae
Although very abundant on most coral reefs, the various species of *Montipora* are very difficult to identify. Encrusting, branching, and table forms are sometimes present in a single species. It is the second largest group of corals found on Southeast Asian reefs. Habitat is extremely variable ranging from silty coastal lagoons to crystal waters of the outer reef.

1 Corallimorpharian *Amplexidiscus fenestrafer*

2 Zoantharian *Palythoa sp.*

6 Montipora Coral *Montipora sp.*

3 Staghorn Coral *Acropora formosa*

4 Warty Coral *Pocillopora verrucosa*

5 Creamy Coral *Stylophora pistillata*

7 Table Coral (150 cm width)
Acropora hyacinthus; Acroporidae
This species typically grows in colonies that form delicately sculptured, broad flat plates. It is one of the most common corals seen on the upper edge of outer reef slopes and on reef flats. Most colonies are cream-coloured or greenish brown sometimes with blue or pinkish margins, indicating the zone of active growth.

8 Lobate Coral (150 cm height)
Porites lobata; Poritidae
This species forms one of the reef's most impressive structures—massive round or helmet-shaped boulders of encrusted coral. The largest colonies grow to heights of over six metres and are estimated to be more than 800 years old. Colours usually range from cream to pale brown or green, but shallow water forms are sometimes bright blue or purple. It is common in protected waters of lagoons and coastal reefs and the back margins of outer reefs.

9 Flower Coral (8 cm width)
Alveopora fenestrata; Poritidae
This species is very similar to *Goniopora*. Both have polyps on long, fleshy stalks. However, they can easily be distinguished. *Alveopora* has 12 tentacles at the end of each stalked polyp compared to 24 in *Goniopora*. Flower corals occur in a variety of habitats including turbid coastal reefs and clear waters of outer reefs.

10 Anemone Coral (15 cm)
Heliofungia actiniformis; Fungiidae
Members of the family Fungiidae are commonly referred to as mushroom corals due to their characteristic shape. They are the most common type of unattached or free-living corals found on tropical reefs of the Indo-Pacific. In other words a single individual is composed of only one polyp in contrast to the colonial structure of most corals. Another distinctive aspect of these corals is their ability to move. Most can dig themselves out if buried under the sand and many are capable of slow lateral movement.

11 Mushroom Coral (15 cm)
Fungia scutaria; Fungiidae
Variable in colour, this species is usually green or yellowish. It typically has light coloured tentacle tips giving the impression of rice grain-like speckling.

12 Tentacle Coral (30 cm height)
Goniopora sp.; Poritidae
This looks like soft coral, but the stalked tentacles of the colony are attached to a typical hard coral skeleton, although it is usually obscured. An estimated 15–20 species occur in the region and are most common in turbid waters of coastal reefs and lagoons, where wave influence is minimal. *Goniopora* is very aggressive towards other corals. Its long sweeping tentacles tend to destroy other species within reach.

7 Table Coral *Acropora hyacinthus*

8 Lobate Coral *Porites lobata*

12 Tentacle Coral *Goniopora sp.*

9 Flower Coral *Alveopora fenestrata*

10 Anemone Coral *Heliofungia actiniformis*

11 Mushroom Coral *Fungia scutaria*

13 Cellular Coral (300 cm)
Lobophyllia hemprichii; Mussidae
Members of this family are readily identified by the large size of the cell-like individual corallites, and the rough ridge-like margins that separate them from neighbouring corallites. They are either uniform grey, green, or brown or have pale streaks around the cellular margins. This species is abundant on the upper edge of reef slopes and may form extensive colonies.

14 Radial Coral (10 cm)
Symphyllia radians; Mussidae
This coral generally forms round growths covered with prominent meandering ridges. Identification of the three common species in the genus *Symphallia* depends on the width of the depressions between ridges. Those of the Radial coral are considered to be intermediate compared with the other two. The species is very common on outer slopes and fringing reefs.

15 Dome Coral (50 cm height)
Diploastrea heliopora; Faviidae
This is one of the most distinctive coral species in the region. The corallites are densely packed and in the shape of low, rounded mounds giving a very neat uniform surface. Unlike other large corals, the surface of the colonies are seldom penetrated by burrowing molluscs and worms. The notorious Crown-of-thorns starfish is also reluctant to attack it.

16 Crispy Coral (100 cm width)
Oulophyllia crispa; Faviidae
This coral forms massive domes that regularly exceed one metre in diameter. The surface is covered with brown-coloured steep ridges and broad V-shaped valleys. It occurs in a variety of habitats but is most frequently seen in protected lagoons. The polyps are visible only at night and are large and fleshy with prominent white tentacle tips.

17 Symphyllia Coral (30 cm width)
Symphyllia sp.; Mussidae
The species of Symphyllia superficially resemble brain corals *(Platygyra),* but their sculpturing is more robust. They are characterised by the largest valleys of all corals—these are the grooves that run between the prominent raised ridges of the coral skeleton. They form both small colonies as shown here or massive growths measuring several metres across. The soft feeding polyps are extended only at night.

18 Orange Daisy Coral (10 cm width)
Tubastraea faulkneri; Dendrophylliidae
This coral is often sighted by divers while exploring shipwrecks and caves. It is usually seen below 10–15 metres depth. Colonies are generally pink or rose coloured with each corallite forming a low-rimmed tube. The spectacular bright orange polyps have always been a favourite subject of macro-lens photographers.

Opposite: A close-up view of the mouth of a Mushroom coral (*Fungia sp.*) with tentacles extended for feeding.
 Photo by Mike Severns.

13 Cellular Coral *Lobophyllia hemprichii*

14 Radial Coral *Symphyllia radians*

18 Orange Daisy Coral *Tubastraea faulkneri*

15 Dome Coral *Diploastrea heliopora*

16 Crispy Coral *Oulophyllia crispia*

17 Symphyllia Coral *Symphyllia sp.*

19 Anchor Coral (40 cm width)

Euphyllia ancora; Carophylliidae

This is one of the easiest of all corals to identify because of the characteristic anchor-shaped tips of the fleshy polyps. It grows on flat surfaces, usually below 10 metres depth, on coastal reefs and in lagoons. Colonies grow to several metres across and the exposed polyps may completely obscure the underlying skeleton.

20 Bubble Coral (40 cm width)

Plerogyra sinuosa; Caryophylliidae

Colonies resemble a cluster of grey-green grapes. It is one of few corals capable of stinging human skin. The grape-like structures are not the usual soft polyps, but are actually water-filled sacs, or vesicles, that may have a protective function. The true polyps remain hidden during the day and are extended only after dark.

21 Flex Coral (40 cm height)

Favites flexuosa; Faviidae

Faviidae contain the most genera of any coral family and are second only to the Acroporidae in terms of number of species and abundance. These two families are the major contributors to the famous mass coral-spawning events that have been documented for some areas of Southeast Asia and Australia. Flex coral colonies are highly variable, ranging from flat plates to columns up to one metre in height.

22 Laminar Coral (300 cm width)

Turbinaria reniformis; Dendrophylliidae

Turbinaria colonies are extremely variable in shape. So much so that the same species may look entirely different depending on where it is growing. For example, colonies in deeper water receive less light and are differently coloured and often take on a different growth form compared to colonies in well-lit shallow water.

23 Brown Daisy Coral (100 cm height)

Tubastraea micrantha; Dendrophylliidae

This coral is commonly seen on steep outer slopes and in passages through the reef where currents are periodically strong. It forms impressive branching colonies that reach over one metre in height. The brown, thick branches are covered with tube-like protuberances, which contain the fleshy polyps. The polyps are also brown and often extended during the day, but quickly retract if disturbed.

24 Blue Coral (150 cm height)

Heliopora coerulea; Helioporidae

This species is more closely related to soft rather than hard corals. The outer surface has a relatively smooth texture, similar to that of Fire coral. Internally the skeleton is blue—fragments are commonly found on beaches. Colonies usually take the form of vertical plates or columns and are either grey or green with tiny white polyps.

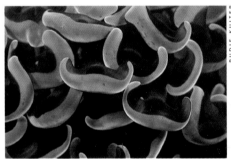

19 Anchor Coral *Euphyllia ancora*

20 Bubble Coral *Plerogyra sinuosa*

24 Blue Coral *Heliopora coerulea*

21 Flex Coral *Favites flexuosa*

22 Laminar Coral *Turbinaria reniformis*

23 Brown Daisy Coral *Tubastraea micrantha*

25 Black Coral (120 cm height)
Antipathes sp.; Antipathidae

The term Black coral is slightly misleading as colonies are often brown, yellow, orange or frosty white. The skeleton is the only black part. It consists of a horn-like material, often fashioned into black coral jewellery. Trees of black coral measuring 2–3 metres in height are most common on steep drop-offs below 20–30 metres depth, but smaller formations are sometimes seen on vertical surfaces or overhangs in only a few metres.

26 Organ Pipe Coral (colony width 12 cm)
Tubipora musica; Clavulariidae

This is another soft coral relative that has a hard skeleton. It gets its name from the peculiar structure of the skeleton which consists of tiers of organ pipe-like tubes. Each tube contains a polyp with eight feather-like tentacles. These are quickly withdrawn at the slightest disturbance. The unusual skeleton is bright red, but is normally obscured by the expanded polyps. Bits of dead organ pipe are commonly seen on rubble beaches.

27 Harp Coral (100 cm height)
Ctenocella pectinata; (Ellisellidae)

This graceful gorgonian coral is distinguished by its wire-like, bright red branches that grow vertically like the strings of a harp. It commonly grows at depths below 10–15 metres in areas exposed to currents, such as outer reef and lagoon passes.

28 Wire Coral (100 cm height)
Hicksonella sp.; Gorgoniidae

This relative of sea fans is most commonly seen in areas where the currents provide good circulation. The branched, bushy colonies reach about one metre in height. The outer covering consists of a soft leathery "bark" which hides a wiry black skeleton. This skeleton is very similar in appearance to that of black coral and the chunky base can be cut and polished for ornamental use.

29 Delicate Sea Whip (180 cm height)
Junceela fragilis; Ellisellidae

Colonies form stalks that occur in isolation or in dense aggregations that give the seascape a ghostly appearance. Individual stalks may reach a length of 3–4 metres. They grow vertically from the sea floor, but often have hooked tips.

30 Gorgonian Fan (250 cm width)
Subergorgia mollis; Subergorgiidae

A huge variety of sea fans occur in the region. They are colonial animals with a hard horny skeleton made of gorgonin. Some species have delicate interwoven branches that efficiently filter or seine microscopic food from the passing currents. They are most common at depths between 10–30 metres. It generally occurs in areas of strong current, thus insuring an abundant supply of planktonic food.

25 Black Coral *Antipathes sp.*

26 Organ Pipe Coral *Tubipora musica*

30 Gorgonian Fan *Subergorgia mollis*

27 Harp Coral *Ctenocella pectinata*

28 Wire Coral *Hicksonella sp.*

29 Delicate Sea Whip *Junceela fragilis*

CORALS

31 Wrinkled Soft Coral (60 cm width)
Sarcophyton trocheliophorum; Alcyoniidae
This is an abundant species commonly seen in sheltered habitats. When fully expanded the polyps impart a furry appearance, but if they are retracted a very smooth, leathery texture results. Individual colonies are up to a metre across. Aggregations sometimes exceed a width of 10 metres and may be closely interlocked, giving the appearance of a single massive colony.

32 Sea Pen (10 cm height)
Virgularia sp.; Virgulariidae
Sea pens are a group of soft corals with a feather-like appearance. Their name is derived from the feather quill pen. They occur only on sandy bottoms, often on slopes with strong currents. A variety of colours, similar to those of *Dendronephthya*, are exhibited. Each pen represents a colony of animals or polyps that arise on limbs perpendicular to the central stem. They filter microscopic food from the currents. If touched the entire pen quickly disappears into the sand.

33 Dendronephthya Soft Coral (16 cm width)
Dendronephthya sp.; Nephtheidae
Dendronephthya is famous for its wide range of colours. Shades of red generally attract the most attention, but blue, yellow, and white colonies are also common. It grows in areas periodically exposed to moderate or strong currents, usually below 5–10 metres depth.

34 Hand Coral (60 cm height)
Xenia sp.; Xeniidae
Hand corals have non-retractile polyps with relatively large conspicuous tentacles and feathery side branches. Unlike most other soft corals, the tentacles rhythmically open and close every few seconds. The polyps resemble miniature outstretched arms with hands opening and closing.

35 Flexible Soft Coral (50 cm width)
Sinularia flexibilis; Alcyoniidae
Colonies have a broad fleshy trunk and numerous finger-like branches. The spindle-shaped scerites which form a primitive sort of supporting skeleton are densely packed in the soft tissue. When the coral dies, this calcareous material contributes a great deal to the overall sediment which is subsequently recycled to build the reef platform.

36 Dendronephthya Soft Coral (6 cm width)
Dendronephthya sp.; Nephtheidae
Viewed close up these magnificent animals present a graphic example of soft coral anatomy. They lack symbiotic algae or zooxanthellae that mask the internal structure of many other species. The embedded calcareous slivers or sclerites that make up the internal skeleton are clearly visible through the semitransparent trunk and branches. Fleshy polyps are grouped in clusters on the branch.

31 Wrinkled Soft Coral *Sarcophyton trocheliophorum*

32 Sea Pen *Virgularia sp.*

36 Dendronephthya Soft Coral *Dendronephthya sp.*

33 Dendronephthya Soft Coral *Dendronephthya sp.*

34 Hand Coral *Xenia sp.*

35 Flexible Soft Coral *Sinularia flexibilis*

Anemones and Jellyfish

Coral Cousins

Anemones, jellyfish, hydrozoans, and hydrocorals are classified in the Phylum Cnidaria, along with hard and soft corals (see previous section). They are treated separately in this section merely for convenience.

Anemones (actinarians)

An anemone is nothing more than an overgrown coral polyp that lacks a hard skeleton. Structurally it is a very simple animal, basically nothing more than a fleshy cylinder that is closed at both ends. The lower end is called the pedal disc and is used for attachment to coral branches or rocks, which are often buried in sediment. The opposite end is the oral disc and has a centrally located mouth that is surrounded by hollow tentacles. The tentacles are usually shades of green or brown due to the presence of symbiotic algae or zooxanthellae. The outer walls of the anemone's body is called the column. Some species have warty projections known as verrucae on the column. They are adhesive, thus useful in keeping the anemone anchored to the bottom.

Most anemones are well hidden in cracks and crevices, or underneath rocks and dead coral slabs. The most conspicuous species seen out in the open are those found with the colourful clownfishes. Other fishes are stung by the anemone's tentacles, but clownfishes have special chemicals in their mucus coat which prevent the discharge of the nematocysts.

Jellyfish

Typical jellyfish have a basic umbrella or bell shape. They can be differentiated from similar appearing hydromedusas, for example the Portuguese-man-of-war, by the scalloped margin of the bell and they lack a membranous fold or velum around the inner margin of the bell. Tentacular arms, densely packed with stinging cells dangle from the edge of the bell. A central, downward projecting tube contains the mouth or manubrium.

Rhizostome jellys are often seen in the vicinity of reefs. They have a thick, fluffy frill between the upper bell and lower tentacles. The central frill actually contains numerous mouths that are used to trap small organisms from the surrounding water. Some types of jellys may harbour small fishes among their tentacles.

The fish utilise the jelly in much the same manner that clownfishes use their anemone hosts. Evidently the fishes are not stung by the jelly and therefore obtain protection from predators during the vulnerable youthful stages of their life cycle.

Hydrozoans and Hydrocorals

Hydrozoans are a group that includes organisms of very dissimilar appearances and sizes. For example, feathery hydroids, the jellyfish-like Portuguese-man-of-war and rock-hard fire coral are all types of hydro-

zoans. Colonial hydroids are the most common type encountered on coral reefs. Each colony has the appearance of a feather, with a central stalk and many side branches. Tiny polyps with cnidarian tentacles occur on the branches. Some polyps are specialised for feeding and armed with nematocysts. Other polyps function as reproductive organs. The feeding polyps screen the passing current for microscopic organisms and organic debris and this food is then shared among the polyps by an interconnecting digestive tube.

Hydrocorals include both the delicate and richly coloured "lace" (stylasterine) and "fire" (milleporine) corals. The latter can cause a burning sensation on the skin. Although they have a well developed, very brittle skeleton composed of calcium carbonate, hydrocorals are more closely related to hydroids than to true corals.

Below (top): The Upside-down jellyfish *(Cassiopea sp.)* inhabits still waters of bays and estuaries. This is its normal swimming position.

Below (bottom): Tube anemones *(Cerianthus sp.)* have long, delicate feeding tentacles which can be quickly retracted into a leathery stalk.

Living Together: Anemone Symbiosis

The term "symbiosis" means "living together." Many examples occur in the coral reef but the classic example is the relationship of stinging anemones with certain damselfishes, mainly in the genus *Amphiprion*.

There are 10 species of host anemones in the world and all are found in Southeast Asia. Half of the world's 28 species of anemonefishes also occur in the region.

Both partners benefit from their mutual association. The fish depends on the anemone for protection, and preens the tentacles keeping them in a healthy condition. It is a popular misconception that the fish actually feeds its anemone scraps of food. This behaviour does occur in captivity, but is rarely seen in the wild. Anemones capture their own microscopic food.

How are the fishes able to live amongst the stinging tentacles of its host? Contrary to popular belief, the fish is not really immune to the stings. A combination of its unusual swimming behaviour and special chemicals in the mucus coat prevent the stinging cells or nematocysts from firing. This so-called immunity is acquired over a period of several hours when the tiny postlarva settles onto the reef following a brief oceanic or pelagic stage. If the young fish is lucky enough to find an anemone before it is devoured an acclimatisation process follows in which it makes gradual contact with the tentacles. Eventually a chemical change occurs in the mucus coat of the fish and the stinging cells are no longer discharged.

Anemonefishes are never found without their host. The tentacles, which severely sting all trespassers, offer a safe haven. A few species of anemonefish actually enter the mouth of the anemone for brief periods. Anemones, which appear quite healthy, are sometimes encountered without fishes, and are therefore less dependent on the association than their fish partners.

MARK STRICKLAND

1 Tube Anemone (30 cm height)
Cerianthus sp.; Ceranthidae

Tube anemones have their stalk or column encased in a leathery tube. The long, delicate tentacles are fully extended when feeding. Unlike most types of anemones, they have a separate anal pore. Their habitat consists of sand bottoms that are periodically exposed to strong currents which bring a supply of planktonic food.

2 Lace Coral (25 m)
Stylaster sp.; Stylasteridae

This is another organism that resembles a hard coral but is actually a jellyfish relative. It forms beautiful, branching fans, adorning the ceilings of caves and over-hangs. Unlike most types of coral, it retains its brilliant pink, red or purple colour after the polyps die.

3 Fire Coral (150 cm width)
Millepora sp.; Milleporidae

This coral is aptly named as its sting causes a burning sensation. Although it has the appearance of a hard coral, it is actually more closely related to jellyfishes. It is a colonial animal whose tiny polyps are embedded in a limestone skeleton. The potent stinging cells cause a painful rash on contact with the skin. There are several species of fire coral and sometimes a single species has several growth forms. It is best recognised on the basis of its mustard yellow colour and fine texture.

4 Upside-down Jellyfish (7 cm disc diameter)
Cassiopea sp.; Cassiopeidae

This is the most commonly encountered jellyfish in coral reef habitats. It has the unusual habit of resting on the bottom with its arms extended upwards. The arms contain live algal cells or zooxanthellae within its tissues which depend on sunlight to produce nutrients that are leaked to the jellyfish. Therefore, the upside down posture provides maximum exposure to the sun.

5 Stinging Hydrozoan (35 cm height)
Aglaophenia cupressina; Plumulariidae

This organism has a plant-like appearance, but is actually a colonial cnidarian. It is commonly attached to rocks and corals, sometimes forming vast gardens. In spite of its harmless appearance, it is equipped with typical cnidarian stinging cells or nematocysts. The initial pain from the sting is short-lived, but usually results in itchy skin welts that persist for several days.

6 Box Jellyfish or Sea Wasp (45 cm length)
Chironex fleckerii; Chirodropidae

This is one of the most dangerous animals in the sea. The pain from its sting is excruciating and may cause death. Fortunately, it is seldom encountered on coral reefs. The preferred habitat is coastal beaches and sheltered inlets. Care must be taken when wading or swimming in these areas. The best method of prevention is to completely cover up with a wet suit of lycra or rubber.

Opposite: Clown anemonefish *(Amphiprion percula)* are intimately bound to their invertebrate host, which in this case is the Magnificent sea anemone *(Heteractis magnifica)*.

1 Tube Anemone *Cerianthus sp.*

2 Lace Coral *Stylaster sp.*

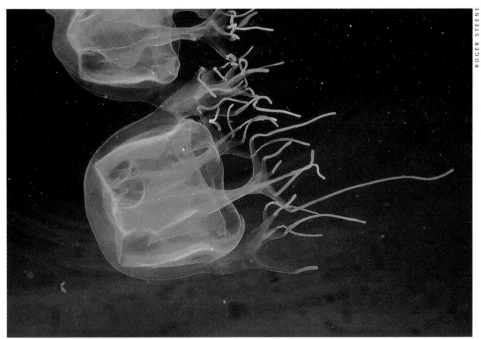

6 Box Jellyfish or Sea Wasp *Chironex fleckerii*

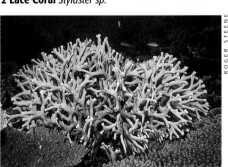

3 Fire Coral *Millepora sp.*

4 Upside-down Jellyfish *Cassiopea sp.*

5 Stinging Hydrozoan *Aglaophenia cupressina*

7 Leathery Sea Anemone (50 cm disc width)
Heteractis crispa; Stichodactylidae

The slender, pointed tentacles and leathery column with adhesive "warts" are distinctive features of this anemone. Tentacles are usually whitish, grey, or brown, rarely violet or bright green. They sometimes have blue or mauve tips. It occurs on outer reefs and lagoons where there are periodic strong currents. The site of attachment is either in sand or rubble or dead portions of branching coral.

8 Mertens' Sea Anemone (60 cm disc width)
Stichodactyla mertensii; Stichodactylidae

This is the largest species of anemone seen on coral reefs, growing to a diameter of one metre or more. It thrives in clear water and strong currents of outer reefs. Tentacles are normally short, although a few long tentacles are present at the centre of the disk. Tentacle colour is usually brown or greenish. The column has longitudinal rows of adhesive red or orange "warts".

9 Bulb-tentacle Sea Anemone (tentacle length 10 cm)
Entacmaea quadricolor, Actiniidae

Opposite: Pink anemone-fish *(Amphiprion perideraion)* seek shelter among the tentacles of their anemone host *(Heteractis magnifica)*. Photo by Jones/Shimlock.

Two distinct growth forms are seen in this species. The first occurs in colonies composed of many individuals and may appear as one giant anemone growing amongst the bases of live or dead corals. The other is a solitary form that may reach nearly one metre in diameter and has very long tentacles. Both forms have greenish-brown tentacles, which often have a distinctive nipple shape.

10 Haddon's Sea Anemone (50 cm)
Stichodactyla haddoni; Stichodactylidae

This anemone is frequently overlooked due to its preference for featureless sand bottoms. The exposed disk is usually grey, brown or green in colour, sometimes with pale streaks. The tentacles are very short. Its usual fish partners are *Amphiprion polymnus* and *A. sebae.*

11 Gigantic Sea Anemone (40 cm)
Stichodactyla gigantea; Stichodactylidae

This anemone is usually found close to shore in shallow, sandy areas. It is sometimes seen in pools on exposed reef flats during low tide. It has a very folded upper surface with relatively short brown tentacles. Usually found with *Amphiprion ocellaris* and *A. percula.*

12 Magnificent Sea Anemone (40 cm height)
Heteractis magnifica; Stichodactylidae

This is the most spectacular of the large tropical sea anemones. It is usually seen in clear water on the edge of outer slopes or in lagoons where currents are periodically strong. The sides of the lower portion of the animal, known as the column, are variable in colour, ranging from brown to red, lavender and blue. The greenish-brown tentacles are long and slender with rounded tips.

7 Leathery Sea Anemone *Heteractis crispa*

8 Mertens' Sea Anemone *Stichodactyla mertensii*

12 Magnificent Sea Anemone *Heteractis magnifica*

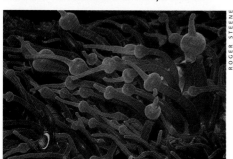

9 Bulb-tentacle Sea Anemone *Entacmaea quadricolor*

10 Haddon's Sea Anemone *Stichodactyla haddoni*

11 Gigantic Sea Anemone *Stichodactyla gigantea*

Worms

Small but Important Reef Inhabitants

Unfortunately the name worm invariably conjures a vision of the not so glamorous earthworm. However, on coral reefs, worms are among the most interesting inhabitants. The multi-coloured feather duster or Christmas tree worms also rate among the reef's most beautiful animals. Worms is actually a catch-all for a diversity of marine creatures belonging to at least 17 major animal categories or phyla. Most are seldom, if ever, noticed by divers. Therefore our coverage in this section is limited to only three of the most conspicuous: flatworms, polychaetes and ribbon worms.

Although not always obvious to divers' eyes, coral reefs are riddled with worms. They live in a variety of micro habitats, particularly in the myriad crevices and fissures. A scientific study on the Great Barrier Reef revealed that a single small coral head harboured 1,400 individual worms belonging to more than 100 species! Many species are burrowers that live either in the dead, basement portion of the reef, or under the surface on silt, sand and rubble bottoms.

Boring worms destroy huge quantities of living coral and their burrows form an endless labyrinth in the reef's foundation. Worms physically pulverise the coral with their horny teeth. There is also evidence that they attack it chemically by dissolving its calcium carbonate matrix. In spite of their destructive tendencies, worms are an integral part of the reef building process. They help reduce coral skeletons to rubble and fine sediment, which is eventually recycled to form a foundation for continued coral growth.

Segmented worms belonging to the Phylum Annelida are familiar to divers. The best known terrestrial equivalent is undoubtedly the earthworm. The most obvious types found on coral reefs belong to a group known as polychaetes which have a variety of shapes and life styles. Some are mobile animals seen crawling over the bottom or on the underside of dead coral slabs and boulders. Others are sedentary and unrecognisable as worms. Many of these build tubes which are produced from mucus secretions, sand grains or mud. The highly ornate Christmas tree and feather duster worms are the most common examples of tube builders.

Sexes are usually separate in polychaete worms, although both sexes are present in several species, and a few others can change sex. Eggs and sperm are either released into the surrounding water or deposited inside the home tube, except in a few species where the eggs are kept on the female's body. The eggs hatch into free-swimming larvae called trochophores. These eventually settle on to the reef after a variable period, frequently of 2–3 weeks duration.

Flatworms belong to the Phylum Platyhelminthes. Included in this group are the liver flukes and tapeworms. These notorious human parasites and the common polyclad flatworms found on coral reefs are miles apart in general appearance. The latter have a much flattened, oval body and colourful patterns. Most species are under about eight centimetres in length. Their locomotion is achieved by sliding across a self-secreted mat of mucus. The actual movement is powered by numerous microscopic bristles. Flatworms can regenerate an entire new animal from a detached fragment but can also reproduce sexually.

Nemerteans or round worms occur in a variety of marine, estuarine, freshwater and terrestrial habitats. They are small and slender and typically have a cylindrical shape. Coral reef species often exhibit very bright colour patterns. They are most commonly found under dead coral slabs or rocks, but also occur in caves and crevices or amongst seaweed. Worldwide the phylum contains approximately 1,200 species yet relatively few are known from Southeast Asian coral reefs.

Below: Scarlet-fringed flatworm *(Pseudoceros ferrugineus)*, about 4 cm long.

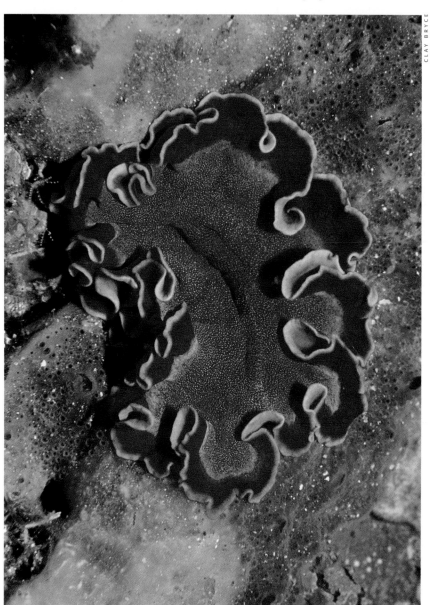

CLAY BRYCE

1. Striped Ribbon Worm (80 cm)
Baseodiscus hemprichii; Phylum Nemertea

Ribbon worms are common on coral reefs. The elongate body is flattened and unsegmented. One of their most notable features is an extendible proboscis. Most species are very small, usually 1–3 cm, but a few reach 5–10 cm or slightly larger. Although many are brightly coloured, they are generally cryptic.

2 Fan Tube Worm (5 cm)
Sabella sp.; Phylum Annelida, Sabellidae.

Polychaete worms in the family Sabellidae are among the reef's most beautiful animals. They live permanently in tubes, which are wedged in coral or buried in the bottom. The graceful feeding tentacles trap small food particles and envelop them in a sticky mucus, conveying them to the mouth. If the tentacles are bitten off by a predatory fish, the worm is unable to feed, but they regenerate in a few days. The light sensitive tentacle tips can detect shadows and cause the tentacles to retract instantly.

3 Bristle Worm (6 cm)
Chloeia sp.; Phylum Annelida, Amphionidae

This polychaete, and similar appearing worms, should never be handled with bare hands as they may have poison glands associated with the bristles which are impossible to remove and cause painful blisters which can persist for a week or more.

4 Spaghetti Worm (tentacles 100 cm long)
Reteterebella sp.; Phylum Annelida, Terebellidae

These white string-like objects are connected to a worm that lives in a tube manufactured from sediments. They function as feeding tentacles, conveying tiny bundles of fine sediment to the worm, which lives well hidden deep below the surface. The strands reach up to a metre or more in length but are quickly retracted if touched.

5 Polyclad Flatworm (3 cm) *Pseudoceros dimidiatus*
Phylum Platyhelminthes, Pseudoceratidae

At first glance polyclads can easily be mistaken for nudibranchs. However, the body is flatter, more lightly constructed, and lacks external gills or other projections evident in most nudibranchs. Flatworms may possibly mimic nudibranchs, which are often toxic and therefore avoided by most predators.

6 Christmas Tree Worm (3 cm)
Spirobranchus giganteus; Phylum Annelida, Serpulidae

Young worms settle on coral heads and secrete a tube that kills the underlying polyps. New coral growth quickly surrounds the tube. Meanwhile the worm occupant secretes additional tube material to keep pace with the coral. The worm lives permanently in its tube. Only the brightly coloured, feather-like feeding tentacles, used to snare planktonic organisms, protrude from its lair. These appendages are light and pressure sensitive and quickly withdraw into the tube.

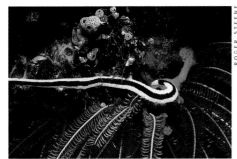
1 Striped Ribbon Worm *Baseodiscus hemprichii*

2 Fan Tube Worm *Sabella sp.*

3 Bristle Worm *Chloeia sp.*

6 Christmas Tree Worm *Spirobranchus giganteus*

4 Spaghetti Worm *Reteterebella sp.*

5 Polyclad Flatworm *Pseudoceros dimidiatus*

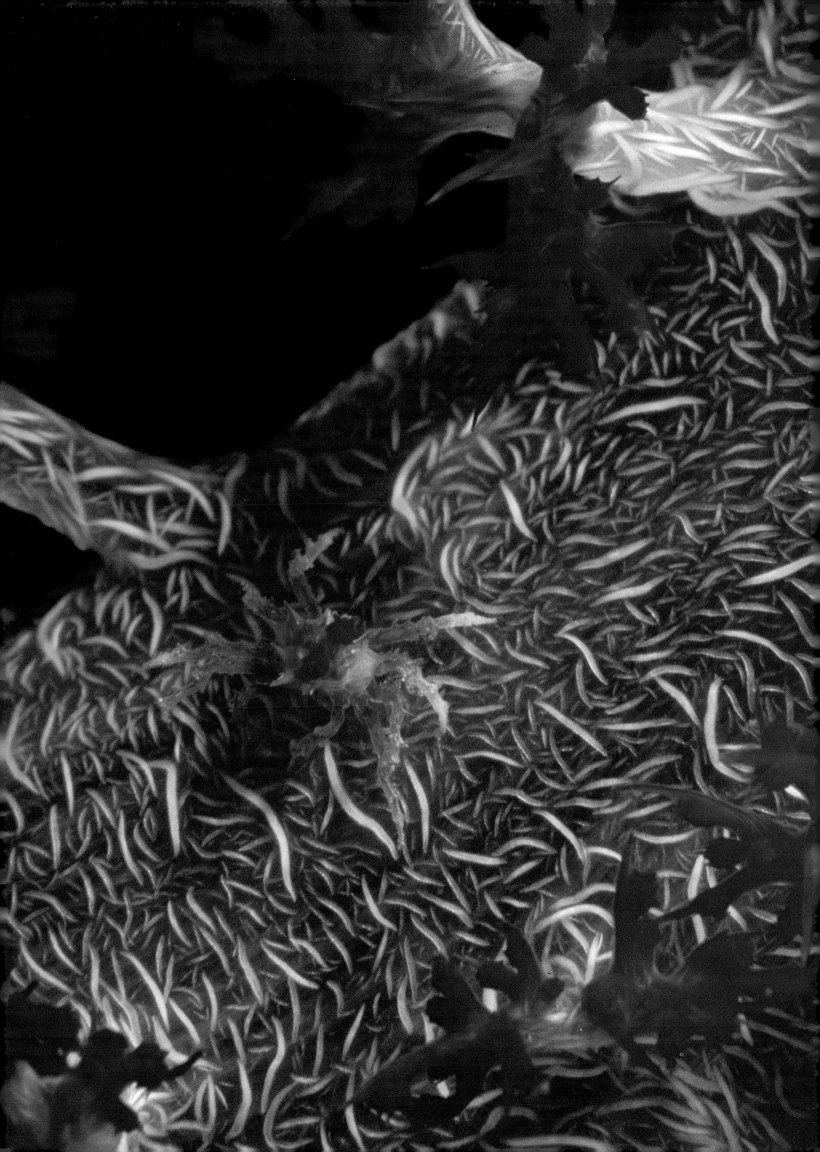

Crustaceans
One of the Reef's Most Dominant Groups

Crustaceans are one of the coral reef's most dominant animal groups. However, due to the tiny size and cryptic habits of many species, it is easy to grossly underestimate their impact. The group is incredibly diverse with regards to size, shapes, colours and life style. Not only does it include well known representatives such as lobsters, shrimps and crabs but also an incredible array of microscopic forms.

Crustaceans are now considered to form a major category or Phylum. They were formerly included in the closely related Phylum Arthropoda, the largest of all animal groups, which contains the largely terrestrial insects, spiders, millipedes and centipedes. Although sharing many features, crustaceans differ from arthropods in having two pairs of antennae instead of a single pair. Over 40,000 species of crustaceans are presently known and hundreds of new species still await discovery.

Besides the obvious types encountered while diving, there is a vast unseen world of microscopic crustaceans. They form a considerable portion of the living soup-like plankton that is of major importance in the food chain that sustains coral reefs. Legions of tiny crustaceans also live in cracks and crevices of dead and live corals, amongst seaweeds and seagrasses, or even between sand grains. Still others form symbiotic relations with a host of diverse organisms, especially algae, sponges, cnidarians, molluscs, echinoderms and sea squirts.

The crustacean body is composed of segments, although they may be hidden by the hard outer crust. There are two major sections—front and back—referred to as the cephalothorax and abdomen respectively. The covering of the head and body is known as the carapace. Another typical feature is the jointed limbs with internal muscular attachment, capable of movement in all directions. These serve a variety of functions that include locomotion, touch and chemical reception, respiration and feeding. The most conspicuous of these are the walking legs, antennae and pincer-like claws or nippers typical of crabs, shrimps and lobsters. The mouth is surrounded by three pairs of appendages specially adapted for feeding: two pairs of maxillae and a pair of mandibles.

One interesting feature of these creatures, is that many crustaceans are able to amputate an injured limb or one that is caught by a predator or other obstacle. There is a special breakage point at the base of each leg where self-amputation can be performed by muscular action and a diaphragm structure prevents blood loss. The lost limbs can then be regenerated.

A pair of compound eyes are present, at least in the adult stage, of most species. Frequently they are placed on the end of movable stalks. The eye of many crustaceans such as crabs and shrimps is composed of cylindrical light-receptors or ommatidia. The number of these microscopic rods is extremely variable depending on the species. Up to 14,000 are present in each eye of the north Atlantic lobster (*Homarus*). Experiments on several types of crustaceans reveal their eyes are able to distinguish size, form and colour, although the total image is apparently crude compared to that formed by the human eye.

Crustaceans are further typified by the presence of a hard external skeleton or cuticle, which is strengthened by calcium salts. Because growth is nearly continuous throughout the life cycle, the animal periodically outgrows it outer coat. Therefore the cuticle is moulted and replaced at regular intervals. Young, fast-growing animals moult at intervals of a few days or weeks, but the period between moults becomes longer with increased age. Before moulting calcium salts are partly absorbed from the cuticle and digested. They are eventually reused to harden the new cuticle. In addition, the old cuticle is sometimes eaten for its calcium salts and organic material. The new coat is initially soft and wrinkled. At this stage, the animal is particularly vulnerable to predators and seeks the shelter of a hiding place for several days until the new shell hardens.

Decapods

The majority of crustaceans seen on coral reefs are known as decapods, Latin for 10 legs. Shrimps (often called prawns), lobsters and crabs are all prominent members of this group. For the most part they are secretive creatures that remain hidden for long periods in burrows or crevices, or under dead coral slabs. They form an integral part of the food chain, being actively hunted by larger predators, mainly fishes. They have evolved a largely inconspicuous life style in response to this pressure. Generally the best time to observe crustaceans is at night when many species emerge from

Opposite: This tiny Spider crab *(Hoplophrys oatesii)* lives on the surface of *Dendronephthya* soft corals. Photo by Jones/Shimlock. ***Below:*** The ever alert Mantis shrimp *(Odontodactyllus scyllarus)* is a voracious predator of crabs, shrimps, gastropods, worms and fishes.

CRUSTACEANS

their retreats to feed. There are relatively few bottom foraging predators at this time but there is still an element of risk. Crab-feeding scorpionfishes (*Dendrochirus* and *Pterois*) are good examples of these nocturnally active predators.

The sexes of decapods and most other crustaceans are separate. The following aquarium observations of the spectacular Harlequin shrimp (see page 33) will serve as a general guide to decapod reproductive patterns. During spawning the male climbs onto the female at right angles to her, hangs underneath her body and transfers its sperm. The animals part after about 30 seconds of copulation and a few hours later about 1,000 eggs appear on the underside of the female's abdomen. Tiny zoea larvae emerge from the eggs in about 20 days.

A free-swimming oceanic larva is typical of most marine crustaceans. The earliest and most basic larval stage is called a nauplius. It is characterised by a single eye and only three sets of appendages. During several successive moults various trunk segments and additional appendages are formed. When the first eight pairs of trunk appendages are free of the carapace, the larva of crabs and shrimps is called a zoea. Once a full complement of functional appendages are present the

young crustacean is known as a postlarva.

The developmental sequence of nauplius, zoea, and postlarva is variable depending on the species and is often modified in the process. For example, most crabs and shrimps pass through the naupliar stage in the egg and hatch out as zoea. Other species bypass both of the early stages and a few skip all three. The postlarval stage of crabs is particularly distinctive and is called a megalops.

Non-decapods

The remaining crustacean groups are sometimes lumped under the heading of non-decapods. They include a variety of diverse organisms but most are seldom noticed while diving on coral reefs. Barnacles, mantis shrimps and a few of the larger isopod fish parasites are among the most conspicuous examples.

Other non-decapods are primarily very small, often microscopic organisms, occurring in great quantities among plankton, on seaweed, on pieces of flotsam, in coral crevices and in the interstices of rubble and sand. This group includes copepods, ostracods, mysids, isopods and amphipods. It forms a vital cog in the food chain, feeding on diatoms and other single-celled plants. These animals, in turn, form the main food for many invertebrates and fishes.

Symbiotic Relationships

Shrimps and crabs form symbiotic associations with a large variety of plants and invertebrates, including sponges, anemones, corals, alcyonarians, molluscs, starfish, feather stars, holothurians and sea urchins. The basis of the association depends on the partners involved. Shrimps and crabs that live amongst the stinging tentacles of sea anemones receive shelter and may also obtain food that is captured by its host. Crustaceans that live on or within sponges may feed on trapped food particles and may benefit their host by helping to keep the water circulation system free of debris.

Several types of snapping shrimps are commonly seen sharing sandy burrows with small gobiid fishes. The shrimp excavates and continuously maintains the burrow, apparently receiving protection from its fish neighbour acting in the manner of a watch dog. The goby stands guard at the entrance of the burrow, signalling the shrimp, with a flick of its tail, when it is safe to emerge. Experiments indicate that the shrimps have extremely poor vision.

Goby Shrimp *Alpheus randalli* (3.5 cm)

Crinoid Crab *Allogalathea elegans*

Cleaner Shrimp *Lysmata amboinensis*

Urchin Crab *Zebrida adamsii* (1 cm)

5
33

CRUSTACEANSCRUSTACEANS

1 Gooseneck Barnacle (5 cm)
Lepas testudinata; Lepadidae

Barnacles were once thought to belong to a separate group, but detailed studies of their life history and anatomy reveal they are modified crustaceans. The jointed appendages of a crustacean are modified to form feeding cirri. These structures are extended outside the shell and their beating movements create a current that conveys planktonic food organisms to the mouth.

2 Mantis Shrimp (12 cm)
Odontodactylus scyllarus; Squillidae

Mantis shrimps come from an ancient lineage, having split away from the mainstream crustacean evolution some 400 million years ago. These large (up to 30 cm), colourful creatures are often seen shuffling across the bottom. They are voracious predators of other crustaceans, small fishes, as well as molluscs and worms, using powerful claws to smash their victim's shell.

3 Snapping (Pistol) Shrimp (3 cm)
Alpheus sp.; Alpheidae

Snapping shrimps are extremely common in this region. Although relatively few species are sighted, their loud cracking noises are frequently the dominant background sounds heard underwater. The noise is actually made by their distinctive claws, which are unequal in size, with the larger usually swollen or flattened. These shrimps are fairly cryptic and remain hidden in crevices.

4 Harlequin Shrimp (5 cm)
Hymenocera picta; Gnathophyllidae

The elegant Harlequin shrimp is characterised by distinctive colour patterns, flag-like antennae, very broad extensions on the ends of the first walking legs that bear the large pincers, and skirt-like extensions on the lower edge of the abdominal segments. Males are generally smaller than females. The diet consists of sea stars, including the Crown-of-thorns starfish.

5 Bumblebee Shrimp (1 cm)
Gnathophyllum americanum; Gnathophyllidae

Also known as the Zebra shrimp, this boldly patterned yet shy species is found in a variety of reef habitats—under rocks and rubble, in caves, and occasionally under sea urchins—and in depths to about 20 metres. The colour pattern is distinctive but the width and number of black bands is variable. The species occurs worldwide in tropical and warm temperate seas.

6 Fish Lice (2 cm)
Family Cymathoidae

Isopods display the greatest morphological diversity of any crustacean group. They are well represented both on land and in the sea. The parasitic fish lice are perhaps the most visible types seen on coral reefs. These flattened, segmented creatures, up to 3–4 cm in length, are often seen attached just behind the head of cardinalfishes (seen here) and other species.

1 Gooseneck Barnacle *Lepas testudinata*

2 Mantis Shrimp *Odontodactylus scyllarus*

6 Fish Lice Family Cymathoidae

3 Snapping (Pistol) Shrimp *Alpheus sp.*

4 Harlequin Shrimp *Hymenocera picta*

5 Bumblebee Shrimp *Gnathophyllum americanum*

CRUSTACEANS

7 Cleaner Shrimp (4 cm) [Also see box on page 32]
Lysmata amboinensis; Hippolytidae
This is by far the most common of several species of cleaner shrimps occurring on Southeast Asian coral reefs. It is easily recognised by the very long white antennae and red and white stripes along the top of the head and body. The habitat consists of caves and ledges in about 5–40 metres depth. It frequently cleans external parasites from moray eels, groupers and other fishes.

8 Marbled Shrimp (4 cm)
Saron inermis; Hippolytidae
Saron shrimps have a spectacular appearance, but are seldom sighted, due to their shy behaviour and nocturnal habits. They hide in crevices or in coral and rubble during the day, emerging at night to feed. The main features for recognition include a pointed snout or rostrum and a toothed crest or ridge on top of the head.

9 Crinoid Shrimp (2 cm)
Periclimenes amboinensis; Palaemonidae
This small inconspicuous reef inhabitant forms a commensal relationship with crinoids, frequently with Bennett's feather star *(Oxycomanthus bennetti)*. The crinoid ranges in colour from yellow to grey or black and the shrimp's colouration invariably matches its host. Other closely related shrimps in the subfamily Pontoniinae live with a variety of hosts including anemones, corals, sponges, sea squirts and bivalves.

10 Reef Lobster (7 cm)
Enoplometopus debelius; Nephropidae
The colourful members of this genus are recognisable by their large claws, spiny bristles and overall lobster-like appearance. They remain well hidden in dark caves and crevices during the day and even at night are seldom seen. Their cautious movement is usually sideways or backwards and the large distinctive claws are held in a raised defensive position. Aquarium observations indicate relatively sensitive eyesight.

11 Cleaner (Banded) Coral Shrimp (5 cm)
Stenopus hispidus; Stenopodidae
The Cleaner shrimp is one of few coral reef crustaceans occurring in all tropical seas. Pairs are commonly seen in crevices or under ledges. They frequently share this habitat with moray eels and are seen walking on the eel's head and entering its mouth in search of parasites. They also clean a variety of other small fishes. Besides their diet of fish parasites, they eat tiny invertebrates.

12 Hingebeak Shrimp (3 cm)
Rhynchocinetes uritai; Rhynchocinetidae
These shrimps are unique in having a relatively long beak or rostrum that is movably hinged to the front of the carapace. They also have disproportionately large, stalked eyes. This species is very gregarious, sometimes forming aggregations of thousands that literally swarm over the surface of large boulders and coral heads.

7 Cleaner Shrimp *Lysmata amboinensis*

10 Reef Lobster *Enoplometopus debelius*

11 Cleaner Coral Shrimp *Stenopus hispidus*

8 Marbled Shrimp *Saron inermis*

9 Crinoid Shrimp *Periclimenes amboinensis*

12 Hingebeak Shrimp *Rhynchocinetes uritai*

13 Painted Rock Lobster (35 cm)
Panulirus versicolor; Palinuridae
Rock lobsters are frequently seen on Asian menus. They differ from their northern hemisphere cousins, the *Homarus* lobsters, in lacking enlarged claws. This species is distinguished from other rock lobsters in the region by the numerous white spots covering the carapace and especially by the longitudinal white stripes on the walking legs.

14 Slipper Lobster (28 cm)
Parribacus caledonicus; Scyllaridae
Slipper lobsters are characterised by a very flat, knobby carapace with heavily recessed eyes. The second pair of antennae are modified to form a very broad, shield-like structure. During the day this animal shelters deep in reef crevices but emerges at night to feed on molluscs.

15 Coconut or **Robber Crab** (35 cm)
Birgus latero; Coenibitidae
This is the world's largest hermit crab, weighing as much as 5 kg with legs that can span 90 cm. Young individuals with a cephalothorax length under about 6–7 cm inhabit abandoned mollusc shells. They are incredible climbers, negotiating such obstacles as sheer rock walls and tall coconut trees. They feed on a variety of vegetable matter, carrion and even other coconut crabs. Ripening fruits and nuts including coconuts, pandanus fruits and figs, are especially favoured.

16 Anemone Hermit Crab (10 cm)
Dardanus pedunculatus; Diogenidae
The relationship of this crab and its anemone hitch hiker is one of the reef's most fascinating examples of symbiosis. When a suitable anemone is located the crab gently strokes and taps it until the cnidarian relaxes its grip on the bottom. The crab then places the anemone on its shell.

17 Anemone Crab (3 cm)
Neopetrolisthes maculatus.; Porcellanide
Anemone crabs are very shy and rarely seen. The body is typically crab-like, but the last walking legs are very small. The clawed legs are relatively large and the flattened abdomen is folded closely under the body. The species shown here is always found living with large sea anemones.

16 Land Hermit Crab (10 cm)
Coenobita perlatus; Coenobitidae
Hermit crabs are among the most conspicuous crustaceans occurring both above and below the water surface. Land hermits are able to survive by retaining moisture in their gill chambers, allowing the gills to function in a similar manner as lungs. This bright red species is very conspicuous along rocky shores or even a considerable distance inland. Females however, must return to the water's edge to release their eggs which hatch out as planktonic larvae.

13 Painted Rock Lobster *Panulirus versicolor*

16 Anemone Hermit Crab *Dardanus pedunculatus*

17 Anemone Crab *Neopetrolisthes maculatus*

14 Slipper Lobster *Parribacus caledonicus*

15 Coconut or **Robber Crab** *Birgus latero*

18 Land Hermit Crab *Coenobita perlatus*

CRUSTACEANS

19 Sponge Crab (12 cm)
Dromidiopsis edwardsi; Dromiidae

Sponge crabs are named after their habit of carrying around living shelter, usually sponges or sea squirts. The last two pair of walking legs, which are clawed and smaller than the other pairs, are used to grip the sponge. The sponge or sea squirts are carried over the back, like an umbrella, and form an effective camouflage. Its carapace is as long as it is broad, and the upper surface is smooth but often covered in fine hairs.

20 Shore Crab (8 cm)
Percnon plannissimum; Grapsidae

Grapsids are one of the most frequently seen crabs. They live in shallow waters of the inter-tidal zone, where they are seen scurrying over the surface of boulders and rocky bottoms or sheltering in crevices. The body is kept tightly pressed against the bottom.

21 Spotted Box Crab (12 cm)
Calappa philargius; Calappidae

Box crabs of the family Calappidae are encountered, while night diving, on sandy bottoms near reefs. They are also known as shame-faced crabs due to their habit of shielding the "face" by holding the large flattened claws close to the front of the body. This species buries itself in the sand during the day, but emerges at night to actively feed on molluscs. It uses the powerful claws to break open the shells of its prey.

22 Decorator Crab (6 cm)
Camposcia retusa; Majidae

Members of the family Majidae are known as spider crabs. The Decorator crab is a master of disguise, using its claws to adorn itself with algae, sponges, and hydroids, which are held in place by numerous hooked hairs on its back. The characteristic triangular carapace, which tapers towards the front, is also covered with spines or knobs. The camouflage makes it almost impossible to detect these creatures when not moving.

23 Halimeda Crab (3 cm)
Huenia sp.; Majidae

This small spider crab is a true master of disguise. The very flattened, lobed carapace is nearly a perfect imitation of the *Halimeda* algal segments on which it is found. The green to brownish colour also matches the surroundings.

24 Squat Lobster (2 cm)
Petrolisthes sp.; Galatheidae

The tiny squat lobsters look like miniature rock-lobsters. They have a flat thorax with a pronounced snout or rostrum. The elongate abdomen is tucked firmly under the thorax. The pair of clawed legs are long and frequently robust, and the last pair of walking legs are very small. Although relatively common, most are inconspicuous due to their small size, which is usually under 2–3 cm. Some species are symbiotic with crinoids.

19 Sponge Crab *Dromidiopsis edwardsi*

20 Shore Crab *Percnon plannissimum*

24 Squat Lobster *Petrolisthes sp.*

21 Spotted Box Crab *Calappa philargius*

22 Decorator Crab *Camposcia retusa*

23 Halimeda Crab *Huenia sp.*

The **Imperial shrimp** *(Periclimenes imperator)* lives symbiotically with echinoderms and the Spanish dancer nudibranch. This individual was found on the surface of a sea cucumber.

The **Huntsman crab** *(Percnon guinotae)* is common in shallow water along rocky shores.

CRUSTACEANS

25 Soldier Crab (2 cm)
Mictyris sp.; Ocypodidae

Soldier crabs are named for their blue and grey military colours and their habit of moving in mass formation over sandy beaches. The body shape is nearly spherical and the walking legs are positioned in a fashion that lifts the animal well above the ground. The habitat consists of inter-tidal sand flats. The crabs burrow into soft sediments in cork-screw fashion, emerging in large numbers at low tide to feed on detritus. Unlike most crabs that walk sideways, soldier crabs often walk forward.

26 Ornamental Crab (3 cm)
Schizophrys dama; Majidae

This species has similar habitats to the Decorator crab. It uses its claws to fasten bits of algae, sponges and hydroids to the upper surface of its carapace, forming an effective camouflage. This crab and its relatives in the genus *Schizophrys* are also known as sea toads due to their relatively robust body and slow movements.

27 Anemone Swimmer Crab (3 cm)
Lissocarcinus laevis; Portunidae

Tiny swimmer crabs in the genus *Lissocarcinus* have a rounded carapace, boldly marked with white and brown. They often form symbiotic associations with cnidarians and echinoderms. This example is found with sea anemones. The anemone provides the crab with shelter and also shares the food it catches with its partner.

28 Swimmer Crab (6 cm)
Charybdis sp; Portunidae

Swimmer crabs of the family Portunidae have flattened rear legs forming paddles used for swimming. In spite of this adaptation most live on the bottom or on floating objects. They are mainly nocturnal predators of molluscs, crustaceans and occasional fishes. The family includes most crabs eaten in restaurants.

29 Ghost Crab (10 cm)
Ocypode cerathopthalma; Ocypodidae

Ghost crabs are common on sandy beaches, mud flats and in mangroves. They construct tunnels in the bottom and emerge at night to feed on detritus. The distinctive stalked eyes can be elevated or folded into grooves at the front of the carapace. They are further characterised by a special hair-lined cavity on each side of the body between the bases of the third and fourth walking legs. The hairs are used to soak up moisture from damp sand which is passed to the gill chamber. Ghost crabs are extremely quick, having been clocked at speeds up to 1.8 metres per second.

30 Soft Coral Crab (2 cm)
Naxioides taurus; Majidae

Spider crabs frequently live on the surface of soft corals and sea fans. They blend in well with the surroundings, hence are very difficult to detect. The very prolonged snout or rostrum is characteristic of this species.

30 Soft Coral Crab *Naxioides taurus*

29 Ghost Crab *Ocypode cerathopthalma*

25 Soldier Crab *Mictyris sp.*

26 Ornamental Crab *Schizophrys dama*

27 Anemone Swimmer Crab *Lissocarcinus laevis*

28 Swimmer Crab *Charibdis sp.*

The **Blue-spotted hermit crab** *(Dardanus guttatus)*. Hermit crabs must periodically change shells to accomodate their growth.

An unidentified **Spider crab** among the branches of a gorgonian soft coral.

CRUSTACEANS

31 Spanner Crab, also known as Frog Crab (3 cm)
Raninoides serratifrons; Raninidae

Spanner crabs have a characteristic elongate, oval carapace. The walking legs are flattened and the last segments are specially adapted for digging into soft bottoms. The movable claws are bent inwards, thus resembling the end of an adjustable spanner. They are also known as frog crabs due to their squat, robust appearance. It is often seen with its rear buried in the bottom.

32 Coral Crab (3 cm)
Trapezia rufopunctata; Trapeziidae

The small (usually 3–4 cm width) coral-inhabiting crabs in the family Trapeziidae have a relatively flat, roughly pentagonal carapace with a broad, jagged front edge. One or both clawed legs is much larger than the other walking legs. Members of the family are associated with branching corals: *Trapezia* species, for instance, live in the branches of *Pocillopora*.

33 Splendid Reef Crab (15 cm)
Etisus splendidus; Xanthidae

Xanthid crabs are well represented on most coral reefs. They are sometimes called black-fingered crabs because of the dark pincer tips, characteristic of many species. They are rather cumbersome, slow-moving crabs that rely on their massive claws to crush open gastropod shells. The Splendid reef crab lives in rocky crevices. This crab should not be eaten as it is poisonous.

34 Spotted Reef Crab (10 cm)
Carpilius maculatus; Xanthidae

Carpilus crabs are recognised by their smooth, rounded carapace, that is often reddish or marked with distinctive patterns. The clawed legs are extremely robust. This species is immediately identifiable by the large red spots on the top and front part of the carapace.

35 Toxic Reef Crab (5 cm)
Zosimus aeneus; Xanthidae

This small xanthid crab is sometimes encountered by waders and snorkellers on shallow reef flats. Carapace, walking legs and claws are covered with a distinctive pattern of red or brown spots and blotches on a pale brown to cream coloured background. The carapace is sculptured with deep grooves and the walking legs are flattened with prominent crests. The species is toxic and should not be eaten.

36 Boxer Crab (2 cm)
Lybia tessellata; Xanthidae

The Boxer crab has a brightly marked, hexagonal carapace and long, slender walking legs. Its most distinctive feature is the small anemone attached to each claw, giving the appearance of boxing gloves. When threatened the crab raises these gloves in a defensive posture and alternately feigns left and right jabs. Besides serving as a defensive weapon the anemones may also assist in capturing food.

31 Spanner Crab *Raninoides serratifrons*

34 Spotted Reef Crab *Carpilius maculatus*

35 Toxic Reef Crab *Zosimus aeneus*

32 Coral Crab *Trapezia rufopunctata*

33 Splendid Reef Crab *Etisus splendidus*

36 Boxer Crab *Lybia tessellata*

Bryozoans

Often Overlooked "Moss Animals"

Although seldom noticed, the bryozoans, or as their name translates—moss animals—are very common on coral reefs. They are colonial organisms, which to the casual observer might easily be mistaken for seaweed or encrusting sponge. They are found in a variety of habitats: under stones, wedged between corals and sponges, and in particular on the walls of caves or in shady crevices.

Bryozoans are one of the most ancient groups of fossil animals dating back to the early Cambrian period. Today the group contains an estimated 4,000–5,000 mainly marine species that occur worldwide. They are generally found in colonies attached to the bottom. Each individual in the colony is a microscopic animal or zooid living in a protective case of calcareous or membranous material. Colonies are extremely variable in size containing anywhere from just a few to millions of zooids.

Adjacent zooids are connected by gaps or pores in the body walls. Members of the same colony may have quite different functions. Many carry out the important function of feeding, but others lack an internal structure and form joints, branches or roots for attachment to the bottom. Still other zooids are modified for reproduction or may have a protective, supportive or cleaning role.

Ciliated tentacles (appendages fringed with fine hairs) at the anterior end of the feeding zooids create a current that drives planktonic organisms into the mouth. The ingested food is moved through a U-shaped digestive tract by means of cilia and waste products are expelled from the anus. The latter structure is located near the mouth but outside of the circle of tentacles.

The members of the colony react in unison, quickly withdrawing into their cases at the slightest vibration. Growth of an individual colony is achieved in plant-like fashion by budding. Sexual reproduction also occurs. Eggs are fertilised and sometimes brooded within the colony. The larvae are free swimming for a short period before attaching on the bottom where growth of a new colony begins.

Bryozoans are one of the first organisms to colonise a bare rock surface. They are notorious fouling organisms that adhere to various marine constructions and the bottom of moored boats. A multitude of growth forms occur but those most commonly noticed on coral reefs resemble delicate lace, plant-like branches, or rather thin, sponge-like crusts. Colour and growth form may be extremely variable within a single species, and to a large extent are determined by environmental factors such as current, light availability and bottom type.

1 Lacy Bryozoan (12 cm)
Iodictyum sp.; Phidoloporidae
This species is recognised by its ruffled appearance, purple coloration and numerous perforations. It is relatively common at most clear water reef areas in the region, but is frequently overlooked due to its habit of attaching to the underside of rocks or under dark ledges. The preferred depth range is between 5–20 metres. It is a colonial organism that is actually composed of hundreds of tiny animals.

2 Branching Bryozoan (8 cm)
Scrupocellaria; Cabereidae
This species is easily mistaken for red algae. It forms branching colonies that grow on the roof of caves or under ledges. It is commonly seen on outer reef slopes at depths between 10 and 30 metres.

3 Graeffe's Bryozoan (6 cm)
Reteporellina graffei; Phidoloporidae
This bryozoan is very common but seldom noticed because of its habit of growing in shady crevices or under ledges along the face of drop-offs. Colonies are roughly circular with many anastomosing branches and several may be positioned in close proximity. They are attached either directly to rocky surfaces or sometimes to dead gorgonian branches. The depth range varies from about four to 20 metres.

1 Lacy Bryozoan *Iodictyum sp.*   **2 Branching Bryozoan** *Scrupocellaria*   **3 Graeffe's Bryozoan** *Reteporellina graffei*

Molluscs
From Tiny Crawlers to Gigantic Clams

Southeast Asia has an incredibly large and diverse assemblage of reef-dwelling molluscs. No official figures are available but experts estimate an impressive total of 5,000 species. A wide assortment of shells, colourful sea slugs and cephalopods are regularly seen by reef walkers, divers and snorkellers.

It is difficult to define a typical mollusc but the majority have a calcareous shell. The shell offers a measure of protection from voracious mollusc predators of which there are many. It also prevents drying of the delicate internal organs. Of course this is not a problem if the animal is constantly submerged, but is of paramount importance to animals living on shallow reef flats that are left high and dry at low tide. However, the shell is definitely not a foolproof character for recognition. Many molluscs, sea slugs and octopuses for example, either lack a shell completely, or it is in the form of an inconspicuous remnant.

The large muscular foot, typical of most molluscs, is another key feature. It allows these animals to creep or crawl over the bottom or bury into the substrate, thus allowing full utilisation of the environment for feeding and shelter. Some molluscs actually use the foot as a fin for swimming or as a float.

The radula is another key molluscan feature. This horny tongue-like structure is situated just inside the mouth. It is used for tearing or scraping off bits of food that are then pulled into the mouth. Herbivores use the device to efficiently graze algae from hard and soft surfaces. Some carnivorous molluscs have a poison gland connected to the radula and inject venom into prey with a harpoon-like radular tooth.

The sex is separate in most gastropods and bivalves and, depending on the species, fertilisation is either external or internal. Giant clams (and nudibranchs) have both male and female organs in the same individ-

ual, but usually the eggs and sperm are not expelled simultaneously. The spawning of these ponderous creatures is an incredible sight. The erupting cloud of gonadal products is an underwater version of Old Faithful geyser. Most gastropods and cephalopods lay strings or clumps of eggs on the bottom. These are often brightly coloured.

The eggs hatch into trochophore or veliger larval stages. The larvae are transparent and have limited mobility, relying on waves and currents for dispersal to suitable reef habitats. This stage of the life cycle may last several weeks. Some species can prolong the larval stage until a suitable reef habitat is contacted. After settling the young animals transform into miniature adults. Octopuses and some other molluscs short circuit the larval stage and hatch out as miniature adults.

Prosobranch Gastropods

This large category of molluscs includes most of the sea shells that are highly prized by amateur conchologists and shell collectors. Cone shells and cowries are probably the best known examples. An amazing wealth of these animals occurs in a variety of Southeast Asian marine environments. In spite of their abundance, relatively few species are seen in rich coral areas. Among the most productive sites are sandy bottoms (look for trails), rubble, and particularly on the underside of rocks or dead coral slabs. Reef flats that are exposed at extreme low tides also offer good possibilities. Remember to return slabs and rocks to their original position. Failure to do this will result in the unnecessary death of numerous encrusting animals.

Most of the body of a typical gastropod is hidden within its shell, which offers protection from predators. However, the smaller species are not immune to the powerful crushing teeth of triggerfishes and some of the larger wrasses.

Aside from the shell, the only other body part usually seen is the muscular foot. This organ is extended and provides a slow, but steady, form of crawling-type locomotion. The foot also secretes a type of mucus which cuts down friction. When disturbed the foot is completely retracted and in some shells there is a solid trapdoor or operculum that completely covers the aperture.

The inner surface of the shell is lined by an organ called the mantle. Special cells in this structure secrete the calcium carbonate matrix of the shell. In cowries and their relatives the mantle can often be seen temporarily enveloping the outer surface and is responsible for maintaining the shell's brilliant lustre. Prosobranch gastropods are characterised by a univalve shell. This means it is composed of a single unit in contrast to the hinged bivalves.

Below: The Coral scallop (*Pedum spondyloidum*) entrenches itself among living coral. Like other bivalves it is a filter feeder, seining micro-organisms from passing currents. The open valves reveal a delicately coloured mantle.

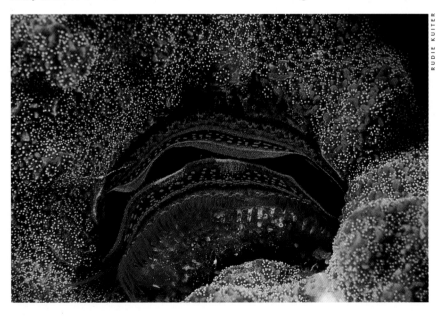

RUDIE KUITER

Opisthobranch Gastropods

These animals are commonly known as sea slugs. Unlike other gastropods, most of the species lack a shell. They are soft-bodied creatures that use their foot-like appendage to slowly crawl over the surface of the reef in search of food. Some species are strictly plant eaters but others are carnivorous. They range in size from large, 50-cm sea hares to tiny microscopic organisms.

The dazzling nudibranchs are easily the best known member of this group. Their scientific name translates as naked gills, an accurate description of their breathing apparatus. In contrast to other molluscs that have gills hidden within their shell or body cavity, those of most nudibranchs form feathery structures on the back. Most species also have a pair of antennae or rhinophores on top of the head.

How do these brightly coloured, slow moving creatures escape predators? The flesh of opishthobranchs is frequently toxic or distasteful due to various chemical secretions. The brilliantly coloured patterns serve to warn enemies of the potential danger. Once a fish inadvertently samples a sour nudibranch it is not likely to forget this experience. In subsequent encounters the nudibranch's bold colour pattern serves to remind the fish of its inedible qualities and thus this animal is ignored.

Nudibranchs and their relatives also utilise camouflage colours very cleverly to escape detection. Species found among seaweeds, sponges and on various cnidarians often effectively blend with the surroundings affording them a large degree of protection.

Bivalves

Bivalves, as their name suggests, are composed of two separate halves. When closed the valves protect the animal from most enemies except some predatory fishes that crunch the entire shell with powerful jaws or some starfish that are able to pry open the valves. Most of the reef species attach themselves to the bottom or burrow into sand, rock, coral or wood. Giant *Tridacna* clam shells are the most familiar example. The two shells are held together by a ligament and strong muscles, which control the opening and closing movements. A pair of openings in the fleshy mantle facilitate water movement through the gill chamber. Bivalves are usually filter feeders. Micro-organisms and other food particles are trapped by mucus on the gills and passed into the mouth.

Cephalopods

Worldwide there are about 650–700 species of cephalopods. They inhabit every conceivable undersea habitat. Many live in the open ocean or in abyssal depths. Relatively few species are encountered on Southeast Asian reefs but at certain localities they are seen on nearly every dive. These animals are renowned for their ability to elude enemies by squirting a inky "smokescreen". The ink is produced by a gland within the fleshy mantle that encloses the other internal organs. It is ejected from a muscular siphon on the lower edge of the mantle. The siphon also provides a form of jet locomotion when water is forcibly expelled from the mantle cavity through its opening. Squids and cuttlefish also have membranous lateral fins along the edge of the mantle that are used for stability, steering and slow swimming.

The name cephalopod literally means head-foot in reference to the two dominant body parts, the foot half being the arms or tentacles. Although they seem unique, these animals have the basic molluscan body plan, but with special modifications. In addition to the eight-legged octopus, the group also contains the 10-legged cuttlefish and squids, and the unusual chambered nautilus.

The octopus and its relatives have the most advanced nervous system of all invertebrate animals. Many of the lower animals are able to detect pressure waves or the difference between shade and sunlight but none have the keen sense of vision enjoyed by the cephalopods. Their eyes are remarkably human-like and accurately register shapes, textures and colours. The keen eyesight and well-developed brain enable them to deftly catch elusive fish prey. In addition, they feed on crabs, shrimps, and in the case of the octopus, bivalve molluscs. Most octopus are harmless, but the Blue-ringed has a toxic bite.

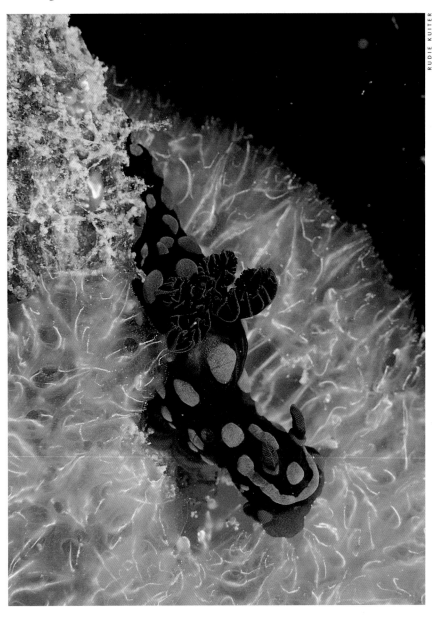

Below: Nudibranchs are opisthobranch gastropods, commonly known as sea slugs. Unlike prosobranch gastropods they usually lack a protective shell. Instead, they rely on toxic secretions for protection from enemies. This is the Green-spotted nudibranch (*Nembrotha kubaryana*).

RUDIE KUITER

MOLLUSCS

1 Rooster-comb Oyster (12 cm)
Lopha folium; Ostreidae

This bivalve is commonly seen in caves, shipwrecks, and other sheltered, shady locations. The distinctive zig-zag shape of the shell margins is unmistakable. Hydroids and gorgonian soft corals are favourite sites for attachment, but it is also seen on rocky surfaces. They are invariably covered with encrusting growth, often bright red sponge or colonial sea squirts.

2 Egg Cowrie (7 cm)
Ovula ovum; Ovulidae

Gastropods in the family Ovulidae have smooth, glossy shells and are very similar to cowries. It is usually seen with the fleshy soft coral, *Sarcophyton*, but also occurs on gorgonian sea fans and various cnidarians. These sedentary animals form a significant part of the egg cowry's diet. The shell is pure white, in vivid contrast to the jet-black mantle. It grows to a length of about 12 cm.

3 Helmet Shell (30 cm)
Cassis cornuta; Cassidae

Helmet shells are popular with collectors and are sometimes fashioned into ornaments. They have a very robust shell that is virtually predator-proof. The habitat consists of extensive patches of clean, white sand. It is frequently buried during the day with only the tip of the shell showing through the sand. It actively hunts at night for food, mainly sea urchins and sand dollars.

4 Partridge Tun (40 cm)
Tonna perdix; Tonnidae

This animal is sometimes seen slowly gliding over rubble or sand bottoms. It feeds exclusively on sea cucumbers, first engulfing the prey then slowly breaking it into small bits within the mouth. It is a slow process, taking up to several hours to ingest the meal. The tun is unable to withdraw into its shell while feeding and is therefore vulnerable.

5 Hairy Trumpet (6 cm)
Cymatium lotorium; Cymatiidae

This shell is often seen in sheltered coastal water and lives under rocks or coral slabs. It has a characteristic trumpet-shell shape with a relatively large aperture and tapers to a point. The shell is characterised by numerous hairy projections. It is a carnivore that consumes sea squirts, tube worms and various echinoderms.

6 Spindle Cowrie (4 cm)
Phenacovolva sp.; Ovulidae

Spindle cowries have a very long, drawn-out shell. They are found among the branches of gorgonians and feed on their polyps. Many of the species are very small and coupled with their excellent camouflage are virtually invisible. It requires an experienced eye to detect these animals as both the shell and mantle usually match the colour and texture of the host gorgonian.

1 Rooster-comb Oyster *Lopha folium*

4 Partridge Tun *Tonna perdix*

5 Hairy Trumpet *Cymatium lotorium*

2 Egg Cowrie *Ovula ovum*

3 Helmet Shell *Cassis cornuta*

6 Spindle Cowrie *Phenacovolva sp.*

7 Triton Trumpet (30 cm)

Charonia tritonis; Ranellidae

The Triton trumpet is one of the few known predators of the Crown-of-thorns starfish, a notorious coral feeder. It was once feared that a reduction in trumpet shell abundance due to shell collectors was a contributing factor in starfish population explosions. However, the trumpet's role in keeping starfish populations in check appears to be over exaggerated.

8 Thorny Oyster (15 cm)

Spondylus varius; Spondylidae

Although this bivalve is extremely well camouflaged it is very conspicuous when the two halves of its shell are open. This is because the mantle forms a beautifully ornate curtain across the entrance to the mantle cavity. It is very sensitive to unusual water movement and snapping shut is a protective response. The thorny, outer surface of the shell provides anchorage on rocky surfaces.

9 Giant Clam (60 cm)

Tridacna crocea; Tridacnidae

Five species of giant clams are commonly seen in our region. The largest grows to a diameter of 1.5 metres, and is reputed to live up to 200 years. They possess microscopic algal cells, zooxanthellae, in their exposed mantles. The algae produce most of the essential food for its host. The mantle also contains light and pressure sensitive spots that cause the shell to quickly close its valves if disturbed.

10 Scorpion Shell (10 cm)

Lambis scorpius; Strombidae

Members of this family are called strombs. They are usually seen in sand or rubble patches on coral reefs. The margin of the shell aperture is widely flared in the adult stage of most species. It has a claw-like, horny operculum that can be used as a lever to upright itself if the shell is flipped onto its back.

11 Common Turban Shell (12 cm)

Tectus niloticus; Trochidae

The underlying shell of this species has a beautiful pearly lustre and is used for making jewellery and buttons. The shell has a distinctive conical shape with a broad, rounded base. When the animal withdraws into its shell, the opening is covered by the operculum, a door-like membrane.

12 Onyx Cowrie (3 cm)

Cypraea onyx; Cypraeidae

The cowrie family needs little introduction. About one half of the nearly 200 known species in the world occur in the SE Asian region. Cowries have a bilobed mantle that completely envelops the shell when fully extended, preserving the brilliant lustre of the shell. Surprisingly little is known about the feeding habits of cowries.

7 Triton Trumpet *Charonia tritonis*

8 Thorny Oyster *Spondylus varius*

9 Giant Clam *Tridacna crocea*

12 Onyx Cowrie *Cypraea onyx*

10 Scorpion Shell *Lambis scorpius*

11 Common Turban Shell *Tectus niloticus*

MOLLUSCS

13 Tiger Cowrie (8 cm)
Cypraea tigris; Cypraeidae
This is one of the largest and most beautifully marked of all cowries. Colouration is extremely variable, but often consists of large, dark-brown spots. It is usually seen in caves or underneath slabs of dead coral but may be out in the open, particularly at night.

14 Murex Shell (8 cm)
Murex sp.; Muricidae
These beautiful shells are delicately sculptured and frequently have spiky projections. The circular aperture is equipped with a horny operculum that seals up the shell when the animal withdraws inside. Murex shells are predators of other mollusc species and barnacles.

15 Olive Shell (4 cm)
Oliva sp.; Olividae
Olives are elongate shells with a very long aperture and a spiral, conical cap. They have a glossy sheen, similar to cowries and are also a favourite of shell collectors. The colour patterns are distinctive but may show considerable variation within a single species. They occur on sand bottoms and are more commonly seen at night. Their diet consists of a variety of other mollusc species.

16 Banded Vexillum (6 cm)
Vexillum taeniatum; Costellariidae
The members of this family are closely related to mitre shells. Both have an elongate, conical shell, but that of the costellariids usually has more pronounced spiral tiers. They inhabit sandy or rocky bottoms in relatively shallow water. Egg cases, containing only a few embryos, are laid on the bottom. There is no larval stage and the embryos hatch as crawling young. They are carnivores, feeding on other molluscs and invertebrates.

17 Marmorated Cone (6 cm)
Conus marmoreus; Conidae
Over 100 species of cone shells are found in the region. They live on or under sandy surfaces or under rocks and coral boulders. They have a very characteristic conical shape with an elongate aperture that is protected by a small, horny operculum. The Marmorated cone has a dangerous sting and should be handled with care.

18 Textile Cone (7 cm)
Conus textile; Conidae
This is one of the most dangerous marine animals due to its potent sting. The tongue-like radula is equipped with harpoon-shaped teeth that contain poisons used to stun and kill small fish prey. The venom is very potent and capable of causing human fatalities. The shell should never be directly handled, nor should it be placed in shirt, pants or diving vest pockets in close proximity to the skin. Other cowries also possess a toxin, but it is usually less powerful and used for paralyzing worms or molluscs.

13 Tiger Cowrie *Cypraea tigris*

14 Murex Shell *Murex sp.*

15 Olive Shell *Oliva sp.*

18 Textile Cone *Conus textile*

16 Banded Vexillum *Vexillum taeniatum*

17 Marmorated Cone *Conus marmoreus*

19 Sacoglossan Sea Slug (3 cm)
Cyerce nigricans; Polybranchiidae
Although not a true nudibranch, this elegant animal completely lacks a shell. It has a pair of double-pronged antennae (rhinophores) on top of the head. Its most remarkable feature is the leafy appendages on the back. These contain extensions of the gut and also glands that secrete a noxious chemical thus discouraging any potential predators. It feeds exclusively on green algae.

20 Pearl Oyster (15 cm)
Pinctada margaritifera; Pteriidae
Several types of pearl oysters are seen in SE Asia. They are commonly attached to rocky surfaces, gorgonians and shipwrecks. Very few natural pearls are found but they are routinely cultured at pearl farms. A tiny nucleus is inserted into the shell around which the shell adds material, eventually forming a pearl. After several years it is ready for harvesting for the jewellery trade.

21 Bubble Shell (2 cm)
Hydatina amplustre; Hydatinidae
This species has a thin, boldly marked shell, but is more closely related to nudibranchs and other sea slugs. It belongs to a large and diverse group of opisthobranchs, the cephalaspideans, that show various stages of shell development. Some families, such as the Hydatinidae, have a well-developed shell, others have an inconspicuous vestigial shell. This species has a large, delicate foot.

22 Tailed Sea Slug (5 cm)
Chelidonura electra; Aglajidae
Tailed sea slugs of the family Aglajidae are closely related to bubble shells. It has a very small, mainly internal shell in the centre of its body. A pair of large flaps, the parapodia, fold up on the sides to the centre of the back and extend from the rear end of the animal, forming a pair of tails, with one flap longer than the other.

23 Forsskal's Slug (10 cm)
Pleurobranchus forskali; Pleurobranchidae
This animal belongs to a group known as the side-gilled sea slugs because of the elongated, plumed gill that is located on the right side of the body. Its colour and texture closely matches the sponge upon which it feeds. Although seemingly vulnerable to predators it protects itself by secreting a foul-tasting acid substance.

24 Sea Hare (30 cm)
Dolabella auricularia; Aplysiidae
The sea hares are among the largest of the opisthobranchs. Some species reach a length of up to 60 cm. Individuals are hermaphroditic, containing both male and female sex organs. They form breeding aggregations in which all participants except the ones at the front and back act simultaneously as a male to the partner in front and female to the one in back. A single animal may produce as many as 180 million eggs during its short lifespan.

19 Sacoglossan Sea Slug *Cyerce nigricans*

20 Pearl Oyster *Pinctada margaritifera*

21 Bubble Shell *Hydatina amplustre*

24 Sea Hare *Dolabella auricularia*

23 Forsskal's Slug *Pleurobranchus forskali*

22 Tailed Sea Slug *Chelidonura electra*

MOLLUSCS

25 Morose Nudibranch (6 cm)
Tjamba morosa; Polyceridae

Nudibranchs are found in nearly all reef habitats. They occur from tide pools down to the greatest depths penetrated by scuba divers. They are seen both in the open and under rocks and coral slabs. The key to finding them is to look for their favourite foods. The species shown here feeds mainly on bryozoans, but other nudibranchs prefer algae, sponges and cnidarians.

26 Spanish Dancer (25 cm)
Hexabranchus sanguineus; Hexabranchidae

Ever since its discovery more than 150 years ago, the exquisite Spanish dancer nudibranch has fascinated observers. It is usually seen slowly crawling over shallow reefs, but if disturbed it swims off the bottom by rhythmically undulating its body. A wide range of colour patterns are seen. It feeds on sponges, mainly encrusting types.

27 Kunie's Chromodoris (6 cm)
Chromodoris kuniei; Chromodorididae

The colourful chromodorids include some of the most beautiful nudibranchs. Worldwide the family contains an estimated 300 species, at least half of which are found in the SE Asian and Australian regions. The species seen here occurs in rocky areas, usually in about 5–30 metres depth. It is readily identified by the bright purple spots and margin. The species feeds on sponges and can store the sponge's toxic chemicals and use them for its own defence.

28 Blacksnout Kentrodoris (16 cm)
Kentrodoris rubescens; Kentrodorididae

This nudibranch is frequently noticed due to its large size and striking colouration. Individuals may reach 17 cm. It is occasionally seen on coral reefs, usually on sand or rubble bottoms. The dark antennae and rounded snout give the head a mammalian look.

29 Aeolid Nudibranch (12 cm)
Phyllodesmium briareus; Glaucidae

The aeolid nudibranchs are easily recognised by the numerous tentacle-like projections, the cerata, covering the back. They are predators of cnidarians including hydroids, anemones, hard and soft corals. The cnidarian tissue containing potent stinging cells are consumed by the nudibranch, which can then use the stingers for their own protection. These stolen weapons are stored in the tips of the cerata.

30 Magnificent Chromodoris (4 cm)
Chromodoris magnifica; Chromodorididae

This species is recognised by the light and dark blue central area and broad yellow-orange submarginal band. The gill filaments and antennae are bright orange. It feeds on sponges and is often sighted crawling over sponge or rock. Depth range is 3–20 metres.

25 Morose Nudibranch *Tjamba morosa*

28 Blacksnout Kentrodoris *Kentrodoris rubescens*

29 Aeolid Nudibranch *Phyllodesmium briareus*

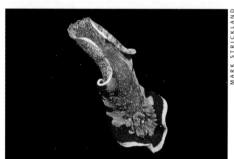

26 Spanish Dancer *Hexabranchus sanguineus*

27 Kunie's Chromodoris *Chromodoris kuniei*

30 Magnificent Chromodoris *Chromodoris magnifica*

31 Celestial Phyllidia (8 cm)
Phyllidia coelestis; Phyllidiidae
Phyllidiids have a tough, leathery body that is moulded on the upper surface into a series of bumps or ridges. These features are usually brightly coloured and greatly aid individual species recognition. The inconspicuous gills are hidden away on the underside of the animal. The members of the family are sponge feeders which can concentrate the sponge's toxins and use them for their own defence.

32 Chambered Nautilus (20 cm)
Nautilus pompilus; Nautilidae
This animal is rarely seen by divers but its shell is frequently found on beaches or floating at sea. The boldly patterned shell provides protection and also serves to regulate the animal's flotation. The nautilus is the most primitive and archaic cephalopod dating back more than 400 million years. Now only a few deep water species survive. The species shown here inhabits depths between 60–250 metres.

33 Blue-ringed Octopus (7 cm)
Hapalochlaena sp.; Octopodidae
This small octopus is very secretive and lives in coral crevices or under rocks. Although it is shy and has an inoffensive appearance it should not be handled under any circumstance. It can deliver an extremely venomous bite that may prove fatal to humans.

34 Broadclub Cuttlefish (12 cm)
Sepia latimanus; Sepiidae
Two species of very large cuttlefish are frequently seen hovering above the bottom on Southeast Asian coral reefs; the Broadclub cuttlefish, shown here, and the tiger-striped Pharaoh cuttlefish. Both are good to eat and keenly sought after by local fishermen. Divers are usually able to approach these animals to within a metre or so.

35 Reef Octopus (40 cm)
Octopus cyanea; Octopodidae
This is the most commonly encountered octopus on coral reefs of the Indo-West Pacific region. Any large octopus that is seen during the day while diving is likely to be this species. It dwells in holes and caverns and like most octopuses can change its colour quickly and dramatically. Most reef octopus species have a very short life span that ranges from about 1–3 years.

36 Mimic Octopus (10 cm)
Octopus sp; Octopodidae
This octopus is one of the world's most amazing marine creatures. It uses its lithe, flexible body and endless colour changes to assume the shape of a whole range of creatures that share its sand-bottom habitat. Anemone, hermit crab, lionfish, flounder and sea shell are all among its "copy-cat" repertoire. This animal has to be seen to be believed.

31 Celestial Phyllidia *Phyllidia coelestis*

34 Broadclub Cuttlefish *Sepia latimanus*

35 Reef Octopus *Octopus cyanea*

32 Chambered Nautilus *Nautilus pompilus*

33 Blue-ringed Octopus *Hapalochlaena sp.*

36 Mimic Octopus *Octopus sp.*

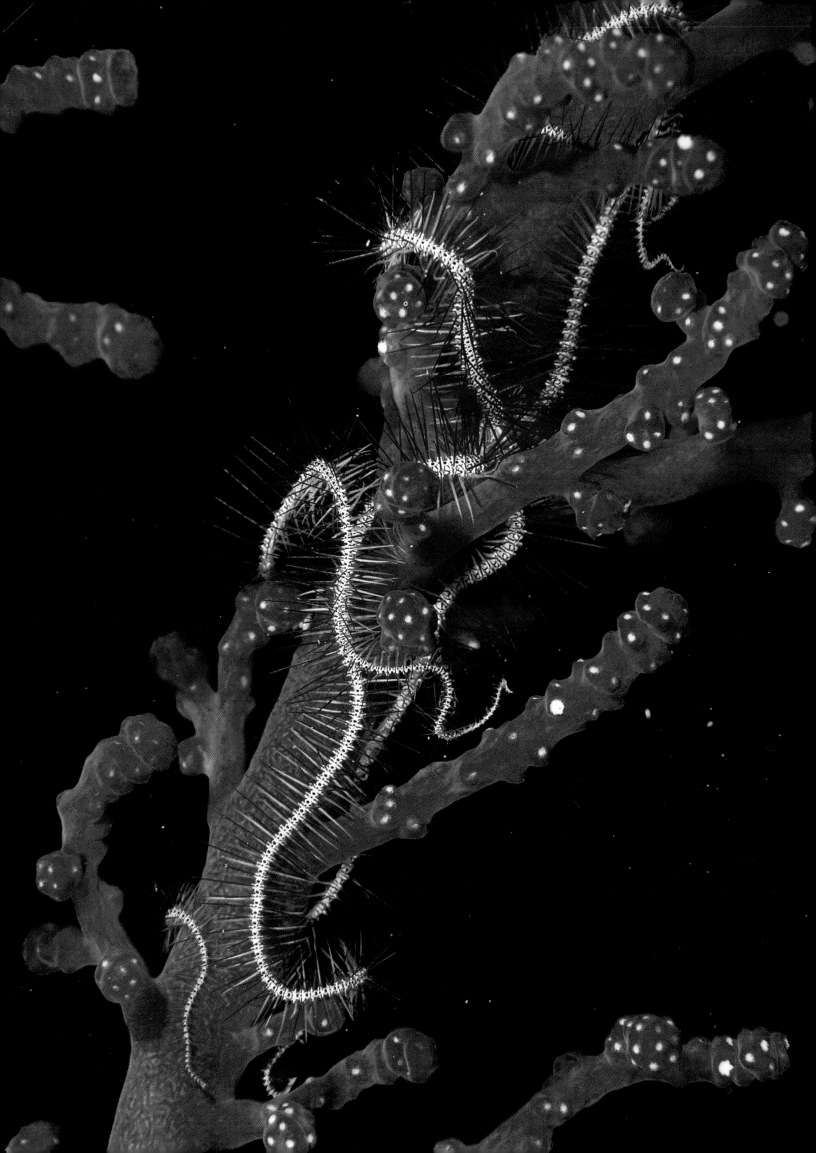

Echinoderms

Amazing Animals with Weird Habits

The Phylum Echinodermata contains some of the reef's most extraordinary creatures. Several species allow fish to reside in their gut cavity, entering and exiting through the anus (see box on page 53). Others are capable of regenerating an entire new animal from a tiny fragment. Still others can disembowel themselves to distract predators. These are but a few examples of their amazing repertoire of behaviour.

An excellent diversity of echinoderms can be viewed on Southeast Asian reefs. Approximately 500 species are found representing the five major divisions: sea stars, brittle stars, feather stars, sea urchins and sea cucumbers. Most readers will be familiar with the five-rayed symmetry of sea stars. This same pentagonal theme is present in other echinoderms as well, although it is not always obvious on the outer surface.

Several other important features are common among echinoderms. They have a skeletal framework composed of calcium carbonate ossicles. In some cases these are close-fitting and form solid plates but in others they are in the form of loosely scattered slivers or spicules. Deserving special mention are the tube-feet, which are easily seen in sea stars on the underside of the arms. These tube-feet are connected to a clever plumbing installation known as the water vascular system. The tube-feet are like miniature, flexible pipettes. When water is pumped into them they respond by con-tracting or retracting, and can also be waved from side to side. Another feature is the pedicellariae. These, pincer-like structures usually arise from a short stalk or base and function in catching food and keeping the surface clear of debris. This is why sea stars never have any encrusting sponge or other growths. These pincers are often very numerous on the outer surface, but are barely visible to the naked eye.

Sexes are separate in echinoderms and fertilisation occurs outside of the body. Individual species form spawning aggregations in which the eggs and sperm are released simultaneously, ensuring a good mix. The eggs and larvae have a brief oceanic stage before the young colonise reefs.

Sea Stars (Asteroidea)

These animals are readily identified by their shape and colour. The tube-feet, which line a conspicuous groove on the underside of each arm, are like tiny suction cups. They are used for locomotion and capturing food. Some species feed on clams and other bivalves. The shell is engulfed by the arms and pressure is applied with the tube-feet until the valves are open. The voracious starfish then extrudes its stomach through the opening and digests its meal. Additional items in the sea star's diet include small crustaceans, worms and other echinoderms.

Opposite: The arms of brittle stars are composed of separate calcareous segments bound together with muscle. This arrangement allows considerable flexibility, which is used to good advantage by the Gorgonian brittle star *(Ophiothrix purpurea).* Photo by Rudie Kuiter. **Below:** This brittle star *(Ophiothrix sp.)* clings to a coral branch with its muscular segments. Brittle stars frequently perch on hard and soft corals to gain access to current-borne planktonic food.

MARK STRICKLAND

ECHINODERMS

Sea stars have amazing powers of regeneration. For example, if attacked by a predatory fish, a complete new animal can grow from an uneaten fragment. Divers often come across these small regenerating fragments, also known as comets. It takes about one year for such a comet to grow into a normally proportioned adult.

Brittle Stars (Ophiuroidea)

Unlike sea stars, the arms of brittle stars are composed of separate calcareous segments held together with muscle. This allows these creatures to traverse the bottom with snake-like movements. They are very shy animals that react negatively to light. Hence, they are seldom seen in the open during the day but prefer shady areas under coral slabs or dark rocky crevices. When molested by humans or predators brittle stars display an amazing escape response. They either fragment themselves or throw off entire arms which continue to wiggle.

As in sea stars, the mouth is located on the underside of the central disc. Some species feed on organic debris that is filtered from sand and mud bottoms. Others construct a mucus net and trap tiny food particles. A few species, including the basket stars, sieve planktonic animals from the currents.

Feather Stars (Crinoidea)

Feather stars are those black sticky things that cling to the legs and elbows of your wet suit after the dive is over. Yes, they are a nuisance at times, but they are also a major contributor to the reef's scenic beauty. These graceful animals, also known as crinoids, are best appreciated at night when their feathery arms are fully extended. Individuals and aggregations adorn exposed surfaces. They are frequently seen clinging to gorgonian fans and whip coral, which allows access to the passing currents.

The body is composed of a small saucer-shaped central disk from which the feathery arms radiate. Many of the reef species have 10 arms or multiples of 10. The arms have numerous short side branches or pinnules. These secrete a mucus slime that traps planktonic organisms. The food particles are conveyed to the mouth by a system of grooves by tube-feet and tiny whip-like flagellae.

A set of shorter, tubular arms, or cirri, also radiate from the central disk. They are used to anchor the crinoid to the bottom or for moving. Some species can also swim short distances by rhythmic flapping of the longer feeding arms.

It's well worth taking a close look among the arms of feather stars. They host an incredible array of symbiotic organisms. These include crustaceans such as shrimps and crabs, and bristle worms, molluscs, brittle stars and even a species of fish. One attentive researcher noted 18 different species of symbiotic animals living on a single feather star!

Below: *The Crown-of-thorns starfish (Acanthaster planci).*

Reefs in Peril?

The notorious coral-eating Crown-of-thorns starfish has been at the centre of hotly contested scientific debates over the past 25 years. It is argued that population explosions of this animal have been triggered by man's interference with natural ecosystems. One theory suggests that the use of pesticides and other chemicals in rivers and creeks running into the sea are somehow responsible for the starfish plagues. On the other hand, it is argued that the starfish explosions are a natural phenomenon that occurs often. Past episodes could have escaped attention because diving on coral reefs is a relatively new endeavour. Drilling samples taken on Australia's Great Barrier Reef seem to support the latter view. Corals killed by starfish leave a characteristic imprint that is evident on samples.

Huge numbers of starfish appear regularly on Southeast Asian reefs and are very damaging. They eat the live polyps, particularly those of *Acropora* species, leaving bleached coral skeletons in their wake. Teams of experienced divers can halt these infestations by painstakingly removing the animals or injecting them with chemicals. Caution has to be exercised when handling these creatures as the spines are toxic and can inflict excruciatingly painful wounds. Even if no action is taken, reefs are highly resilient structures and can recover over a period of several years.

MARK STRICKLAND

Sea Urchins (Echinoidea)

At first glance sea urchins bear little resemblance to starfish. However, they do have a similar five-chambered architecture but it is hidden inside the globular outer shell. Urchins are sometimes referred to as sea eggs or sea pumpkins because of the appearance of the hard shell or test, which is frequently seen on the bottom or washed up on beaches.

The most notable feature of the living animal is the battery of long movable spines. These are attached to the shell by ball and socket joints that have muscular attachments allowing the spines to move in all directions. The spines serve as defensive weapons and can readily puncture human skin. In some species they are very long and hollow with needle-sharp tips and backward projecting barbs. The wounds from this sort of spine are very painful due to a toxin which causes a local numbing sensation, headaches, nausea and faintness. Fortunately, the pain soon subsides but numbness and discolouration of the wound site may persist for weeks. The spines can sometimes be picked out with a needle but if left alone will gradually dissolve, although in rare cases surgical removal may be necessary.

The spines on the lower half of the animal can be used as levers in conjunction with the tube-feet for its walking movement. Most species feed on algal scum and seaweed but they also eat small molluscs and other invertebrates. The mouth is centrally located on the bottom of the animal. It is connected to a very elaborate internal jaw apparatus called Aristotle's lantern.

Sea Cucumbers (Holothuroidea)

Sea cucumbers or holothurians are sausage-shaped animals commonly observed on sand bottoms. The mouth is surrounded by branched tentacles that are connected to the water vascular system and can be opened and closed for feeding. In contrast to other echinoderms, its skin is usually thick and leathery, although some species have a thin, almost transparent skin. The embedded calcareous ossicles or spicules give the skin a gritty texture. One scientist estimated each animal contains up to 20 million spicules.

Most species feed on the rich organic film that coats sandy surfaces. These living "conveyor belts" ingest large amounts of sand as they slowly crawl over the bottom. The edible, organic material is digested as the sand particles pass through the straight, tube-like digestive tract. The processed sand is then expelled from the anus, leaving a characteristic trail on the bottom. Several species spurt sticky, toxic threads from the anus when threatened or roughly handled. Others can expel their internal organs to divert the attention of predators.

The name "sea cucumber" is derived from their shape and widespread use in Asia as a base for soups. Considerable commercial trade is based on several species commonly referred to as trepang or *beche-de-mer*. The live animals are gathered from the reef and dried in the sun or with special ovens. The leathery skin is transformed to a gelatinous, rather tasteless concoction during the cooking process.

Below: Assfish (*Enchelio-phus homei*) with its Leopard sea cucumber host (*Bohadschia argus*).

The Wonderfully Weird World of the Assfish

ROGER STEENE

The large complex bodies of echinoderms offer opportunities for a wide range of symbiotic organisms. There are numerous crab and shrimp species that live on the surface of various starfish, sea urchins, feather stars and sea cucumbers. The most interesting relationship, however, involves a species of fish that lives in the gut cavity of certain sea cucumbers.

Although rarely seen outside its host, the Assfish or Pearlfish, enters and exits via the cucumber's anal opening. It feeds on the gonads and other internal organs of the sea cucumber but wisely does not inflict fatal damage.

Assfish are commonly found in the Leopard sea cucumber (*Bohadschia argus*). Virtually every specimen contains a slender transparent fish, between 5 and 10 centimetres in length, in the gut cavity. A few other species are also known to contain fish, including the Prickly sea cucumber (*Thelanota ananas*) and the Spiny black cucumber (*Stichopus coronatus*). The same fish also inhabits *Culcita* and *Acanthaster* sea stars as well as pearl-oyster shells.

ECHINODERMS

1 Blue Linkia Sea Star (25 cm)
Linckia laevigata; Ophidiasteridae

This is the sea star most frequently sighted by divers. Its bright blue colouration is unmistakable. It grows to 40 cm in diameter and is most common on rubble bottoms in relatively shallow depths. Small comets or regenerating bits are occasionally seen and they can develop into a new starfish in as little as a few weeks.

2 Pin Cushion Sea Star (25 cm)
Culcita novaguineae; Oreasteridae

Only the young are recognisable as obvious sea stars. As the animal matures it become globular with inconspicuous arms. Colouration is usually reddish to shades of brown or may be variegated with light and dark patches. It feeds on corals by exerting its stomach through the mouth and externally digesting the polyps. Pearl fish sometimes live in the gut cavity.

3 New Caledonia Sea Star (20 cm)
Nardoa novaecaledoniae; Ophidiasteridae

Radial symmetry is obvious in most starfish. Most types typically have five arms but higher numbers are found in some species. The body actually consists of five equal segments each containing a duplicate set of various internal organs. The centrally located mouth is situated on the bottom or oral side, while the anus is on the top or aboral surface. This species feeds on the organic-rich film that covers rocky and sandy surfaces.

4 Giant Sea Star (100 cm)
Leiaster leachi; Ophidiasteridae

This species is frequently seen by night divers on steep slopes. It occurs on hard rocky bottoms or patches of coral rubble. Distinctive characteristics include the deep crimson colour and very large size. The arms are relatively smooth.

5 Comb Sea Star (12 cm)
Astropecten polyacanthus; Astropectinidae

Astropecten sea stars have distinctive laterally projecting spines on the edge of the arms, which are well adapted for digging. The various species are voracious predators of sand-living invertebrates such as worms, crustaceans and echinoderms. Their bivalve feeding habits are well documented. The prey's valves are forced open with the tube-feet and the sea star's stomach is extruded into the opening. Digestion proceeds slowly and a large meal requires hours or even days.

6 Red-tipped Sea Star (12 cm)
Fromia monilis; Ophidiasteridae

Sea water is drawn into the water vascular system of sea stars through the small sieve plate or madreporite on the upper surface. It is then conveyed to the individual arms by a series of canals. The sieve plate is evident on this species as a red patch located off to one side on the central disk. Its diet consists of encrusting sponges, but it also consumes organic-rich surface slime.

1 Blue Linkia Sea Star *Linckia laevigata*

2 Pin Cushion Sea Star *Culcita novaguineae*

3 New Caledonia Sea Star *Nardoa novaecaledoniae*

6 Red-tipped Sea Star *Fromia monilis*

4 Giant Sea Star *Leiaster leachi*

5 Comb Sea Star *Astropecten polyacanthus*

7 Granular Sea Star (30 cm)
Choriaster granulatus; Oreasteridae

The arms of this sea star have a distinctive swollen appearance with rounded tips. A series of small, pink or reddish gills protrude through the skin on the upper surface of the animal giving it a granular appearance. The habitat consists of coral reef at depths of 5–30 metres.

8 Horned Sea Star (30 cm)
Protoreaster nodosus; Oreasteridae

This sea star is ornamented with robust horn-like nodules that serve a protective function. It is found on sandy bottoms, frequently among sea grass beds. The depth range is from about 3–20 metres. Aggregations are sometimes seen and probably indicate spawning activities. Colouration is quite variable and includes shades of red, blue, yellow and brown. This species is a surface film feeder, ingesting microbial organisms and algae.

9 Serpent Star (10 cm)
Astrobrachion adhaerens; Asteroschematidae

This unusual brittle star may be difficult to detect because of its cryptic habits and camouflage colouration. The best places to find them are steep drop-offs where there is abundant growth of black coral. The Serpent star lives symbiotically with the coral, wrapping its very long arms around the branches of the coral. The usual segmented structure of the brittle star arm is hidden under a layer of soft skin.

10 Basket Star (80 cm)
Astroboa nuda; Gorgonocephalidae

Unusual in appearance, these animals belong to the brittle star group. When fully spread in a current, the numerous branched arms form an effective fishing net for capturing crustacean and fish larvae. Basket stars are most commonly seen after dark.

11 Gorgonia Brittle Star (20 cm)
Ophiarachnella gorgonia: Ophiodermatidae

Most brittle stars lead a cryptic life to avoid predation by fishes, emerging only at night to feed on crustaceans and small invertebrates.

12. Spiny Brittle Star (25 cm)
Ophithrix nereidina; Ophiotrichidae

Members of the family Ophiotrichidae have hairy or spiny lateral projections on the arms. They are frequently found forming symbiotic associations with sponges and gorgonians. The species shown here is living on a sponge. It feeds on organic debris that accumulates on the surface of its host.

13 Frosty-tipped Feather Star (15 cm)
Himerometra bartschi; Himerometridae

The preferred habitat of this species is locations washed by strong currents such as outer reef slopes and lagoon passages. It is often seen clinging to gorgonian fans or whip corals in order to achieve maximum exposure for

7 Granular Sea Star *Choriaster granulatus*

10 Basket Star *Astroboa nuda*

11 Gorgonia Brittle Star *Ophiarachnella gorgonia*

8 Horned Sea Star *Protoreaster nodosus*

9 Serpent Star *Astrobrachion adhaerens*

12 Spiny Brittle Star *Ophithrix nereidina*

ECHINODERMS

feeding on current-borne organisms. The arms and basal part of the pinnule are black and outer pinnules mainly whitish giving it a distinctive frosty appearance. The arms number about 20–30.

14 Noble Feather Star (25 cm)
Comanthina nobilis; Comasteridae

This brilliant species is a conspicuous inhabitant of outer reef slopes and lagoons. Both arms and pinnules are bright golden yellow. The numerous arms are positioned to form a roughly hemispherical shape. This arrangement results in an efficient filtering mechanism for trapping planktonic food particles.

15 Many-armed Feather Star (25 cm)
Comaster multibrachiatus; Comasteridae

As both the common and scientific names of this feather star suggests it possesses numerous arms, sometimes numbering in excess of 100. It is usually seen on outer slopes at depths between 10 and 35 metres. Spawning takes place throughout the year. The gonads on the lower part of the arms ripen first and there is a progression in development with those near the tips developing last.

16 Burgundy Feather Star (20 cm)
Himerometra robustipinna; Himerometridae

The rich red colouration of the Burgundy feather star is best appreciated on night dives when illuminated with a torch. It perches on outcrops of live and dead coral for maximum exposure to passing currents. Depth range is between about 10–50 metres. The arms are moderately thick and generally number in excess of 10. The bright red side branches or pinnules are numerous and close set.

17 Banded Feather Star (20 cm)
Stephanometra sp.; Mariametridae

This elegant animal is usually seen on night dives. It frequents coral reef areas in moderate to strong currents at depths to about 15–20 metres. Arms are relatively thin and number in excess of 30. The numerous close-set pinnules are distinctly marked with alternating light and dark bands. The white areas tend to be broader on the outer portion of the arms.

18 Bennett's Feather Star (20 cm)
Oxycomanthus bennetti; Comasteridae

This feather star is generally common on most reefs bathed by clear water and periodic strong currents. The normal depth range is between 6–30 metres. The arms, numbering from about 80–120, are moderately thick and colour is extremely variable as can be seen here. Green specimens are also seen. Small symbiotic shrimps, crabs and a species of clingfish are sometimes present on the disc or amongst the arms. The short arms or cirri that secure the animal to the bottom are clearly visible on the yellow individual in this photo.

13 Frosty-tipped Feather Star *Himerometra bartschi*

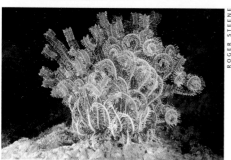

14 Noble Feather Star *Comanthina nobilis*

18 Bennett's Feather Star *Oxycomanthus bennetti*

15 Many-armed Feather Star *Comaster multibrachiatus*

16 Burgundy Feather Star *Himerometra robustipinna*

17 Banded Feather Star *Stephanometra sp.*

19 Variable Sea Urchin (20 cm)
Echinothrix calamaris; Diadematidae
The spines of this urchin vary in colour from pure white to jet black or may have alternating white and black bars. There are also clusters of very fine, needle-sharp, bronze coloured spines sandwiched between the longer spines. A gelatinous sack on the middle of the upper surface discharges faeces. During the day it rests in shady crevices but emerges at night to feed on algae and small invertebrates.

20 Toxic Sea Urchin (18 cm)
Toxopneustes pileolus; Toxopneustidae
The spines are very short on this urchin and it is covered with unusually large pedicellariae. When the animal is alarmed these miniature jaw-like structures open wide giving the urchin an overall floral appearance. Each pedicellaria contains a gland that produces a powerful venom, a small amount of which is capable of killing a rabbit. The scientist who first worked on the toxin nearly lost his life through carelessness.

21 Burrowing Sea Urchin (5 cm)
Echinometra mathaei; Echinometridae
This urchin is common in shallow rocky areas exposed to moderate wave action. It is firmly anchored in holes which it excavates in rock and coral. The strong spines are variable in colour, but they are usually tipped with white.

22 Long Spined Sea Urchin (20 cm)
Diadema setosum; Diadematidae
This is another sea urchin that should be given a wide berth. It has long, sharp spines that contain toxins and can cause very painful wounds. If impaled by the spines the pain can be eased by soaking the wound in hot water. Extreme care should be exercised when entering the water for a night dive. In many areas they form dense aggregations next to the shore after the sun sets.

23 Pencil Urchin (15 cm)
Heterocentrus mammillatus; Echinometridae
This distinctive urchin is often common in shallow water. It has two types of spines. Short flattened spines on the bottom surface are used to anchor or wedge the animal securely in rocky crevices. The slaty pencil-like spines on the upper surface are primarily used for defence. It has few enemies except triggerfishes.

24 Star Urchin (20 cm)
Astropyga radiata; Diadematidae
This sea urchin has clusters of relatively short, needle-like spines separated by longitudinal rows of bright blue spots. It prefers mud, sand or silt bottoms, or seagrass beds. It is a very mobile species that quickly scurries over the bottom with the aid of its spines and tube-feet. It occasionally hosts groups of small cardinalfish of the genus *Siphamia*, which seek protection among the spines.

19 Variable Sea Urchin *Echinothrix calamaris*

20 Toxic Sea Urchin *Toxopneustes pileolus*

24 Star Urchin *Astropyga radiata*

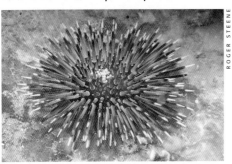
21 Burrowing Sea Urchin *Echinometra mathaei*

22 Long Spined Sea Urchin *Diadema setosum*

23 Pencil Urchin *Heterocentrus mammillatus*

ECHINODERMS

25 Globular Sea Urchin (5 cm)
Mespilia globulus; Temnopleuridae

This small urchin is identified by its short spines which are separated into longitudinal bands by 10 broad spineless areas covered with tiny pedicellariae. It is usually found under coral slabs and rubble, but is seen in the open at night. The upper surface is often covered with algal fragments. It grazes on algae and bottom detritus.

26 Umbrella Sea Urchin (15 cm)
Tripnieustes gratilla; Toxopneustidae

This species has a habit of shading itself with bits of weed, leaves and shell fragments, held in place on the upper surface by the tube-feet. It is usually seen on sand or rubble patches amongst seagrasses or coral in very shallow water. Food items include seagrasses and seaweeds, but detritus and small bottom-living invertebrates are also consumed.

27 Yellow Sea Cucumber (6 cm)
Pentacta lutea; Cucumariidae

This brightly-coloured holothurian is usually seen crawling on the surface of certain sponges in areas exposed to strong currents at depths ranging between about 10 and 40 metres. It is frequently found in groups, sometimes in very large aggregations. The tentacles, which surround the mouth, are fully extended when feeding on plankton.

28 White Patch Sea Cucumber (30 cm)
Actinopyga lecanora; Holothuridae

Members of the genus *Actinopyga* have five calcareous teeth forming a ring around the anal opening. This species is recognised by the distinctive white blotch which also surrounds the anus. It occurs on sand and rubble patches amongst coral reefs, frequently where there is heavy siltation. This species is edible and forms part of the *beche-de-mer* trade in Southeast Asia.

29 Spotted Sea Cucumber (35 cm)
Bohadschia argus; Holothuridae

This species is easily identified by its bold brown spots and is seen either on open sand or in seagrass beds. In some areas every specimen contains an elongate pearlfish that lives symbiotically in the gut cavity. When disturbed the sea cucumber ejects a bundle of sticky threads.

30 Sea Apple (30 cm)
Pseudocolchirus violaceus; Cucumariidae

This is one of the most beautiful of all echinoderms. The globular, relatively soft body comes in a variety of colours and is crowned with flowery tentacles surrounding the mouth. When planktonic food items are trapped its fleshy arm transfers the objects to the mouth. This animal inhabits both flat, silt bottoms and steep dropoffs to depths of about 30 metres. Rarely seen, it is however, abundant on steep walls off southern Komodo and some adjacent islands in Indonesia.

25 Globular Sea Urchin *Mespilia globulus*

28 White Patch Sea Cucumber *Actinopyga lecanora*

29 Spotted Sea Cucumber *Bohadschia argus*

26 Umbrella Sea Urchin *Tripnieustes gratilla*

27 Yellow Sea Cucumber *Pentacta lutea*

30 Sea Apple *Pseudocolchirus violaceus*

31 Graeffe's Sea Cucumber (60 cm)
Bohadschia graeffei; Holothuridae
This is the most spectacular of the sea cucumbers found on the actual coral reef rather than the adjacent sand bottoms frequented by most holothurians. It is usually seen singly, but small aggregations occasionally occur. The normal depth range is about 8–25 metres. The animal crawls slowly over the sea floor with its short, black-coloured feeding tentacles sweeping over the surface in front of the head.

32 Black Sea Cucumber (20 cm)
Holothuria atra; Holothuridae
This conspicuous sea cucumber is very common on shallow reef flats with sand or rubble bottoms, and also frequents seagrass beds. Concentrations can range from about 50 to over 300 animals per 100 square metres. The body and tentacles are entirely black. Small specimens are often coated with sand. A toxic red fluid is released if the outer skin is rubbed.

33 Teatfish Sea Cucumber (35 cm)
Stichopus horrens; Stichopodidae
This species is recognised by the teat-shaped papillae covering the body. The body is very firm but can rapidly disintegrate when the animal is stressed. If the animal is promptly returned to the water and left alone, the process reverses. When attacked, the sea cucumber will shed parts of the body wall in contact with the predator.

34 Prickly Red Sea Cucumber (50 cm)
Thelenota ananas; Stichopodidae
This is the largest or bulkiest species of sea cucumber in the region, reaching a length of 60 cm. The squarish body with numerous projections is unmistakable. These fleshy growths provide shelter for several symbiotic organisms including scale worms, shrimps and brittle stars. Due to its large overall size it is highly prized in the *beche-de-mer* industry which has lead to its extermination in some areas.

35 Serpent Synaptid (150 cm)
Opheodesoma sp.
This species is sometimes seen on sand or rubble patches, frequently amongst seagrasses. It has a very elastic, snake-like body which can exceed two metres when fully relaxed. The outer surface is thin and semi-transparent with a sticky texture due to numerous anchor-shaped spicules embedded in the skin. It feeds on organic debris and algal scum.

36 Sponge Synaptid (10 cm)
Synaptula sp.; Synaptidae
These animals live on large barrel and encrusting sponges. They cling to the outer, exposed surface and are sometimes so densely packed that they completely obscure the sponge. The synaptid obtains microscopic food while the sponge benefits by having its surface kept free of debris so water can circulate through the pores.

31 Graeffe's Sea Cucumber *Bohadschia graeffei*

32 Black Sea Cucumber *Holothuria atra*

33 Teatfish Sea Cucumber *Stichopus horrens*

36 Sponge Synaptid *Synaptula sp.*

34 Prickly Red Sea Cucumber *Thelenota ananas*

35 Serpent Synaptid *Opheodesoma sp.*

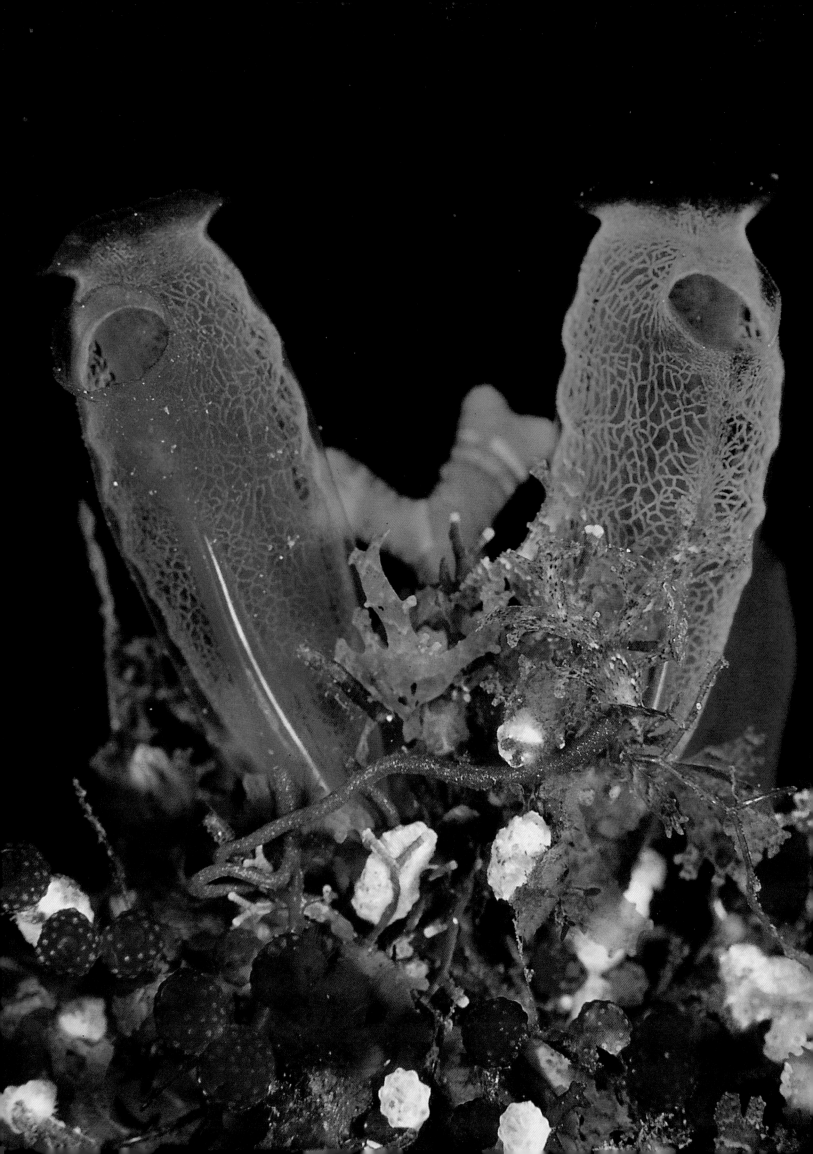

Sea Squirts
Ascidians in all Shapes, Colours and Sizes

Sea squirts or ascidians are common in all marine habitats including coral reefs. They are members of the Phylum Chordata, although superficially they appear to have little in common with their fellow vertebrates. Most of the resemblance occurs in the larval stage when it has a tail that is reinforced with a rod of cells exactly like a vertebrate.

There are more species of ascidians in Southeast Asia than any other region in the world. Every coral reef has a variety of species. The southern Komodo area is one of the best places for viewing and photographing these colourful creatures. Cold upwelling from the deep sea at this locality is conducive to a much richer ascidian fauna than normal.

Ascidians occur as solitary individuals growing up to a size of 20 centimetres, or in colonies composed of numerous, small (sometimes only one to two centimetres) individuals called zooids. The attachment sites are variable, depending on species. The range includes seaweed, sand pebbles, rocks, dead coral fragments, shells and gorgonians. Larger solitary species are sometimes covered with filamentous algae, hydroids, barnacles or even other sea squirts. They may also harbour symbiotic crustaceans that live inside the body cavity.

The basic body plan of a solitary sea squirt is very simple; it is virtually little more than a hollow sack through which water is filtered. The body is surrounded by or embedded in a fibrous material resembling the cellulose of plants. Its texture varies considerably: firm and gelatin-like in some species, delicate and membranous in others, and tough and leathery in still others. Colouration is often bright with all shades of the spectrum. Green-coloured colonies generally harbour symbiotic algae.

Sea squirts come in a variety of shapes. Flat encrusting plates or sheets, stalked or attached directly to the bottom, vases or flasks, lamellae, cones and spheres are all represented. The zooids may be completely or partly embedded in the matrix, or joined to each other at their bases or by creeping stalks. Colonies become increasingly larger due to the ability of the zooids to clone themselves by budding or dividing.

Sea squirts are characterised by a pair of openings: a mouth or incurrent opening, and an excurrent one. The mouth is often directed downwards to avoid clogging with silt or other debris. The excurrent opening, from which waste products are expelled, is usually directed away from the mouth. A continuous current, drawing water into the mouth and out of the smaller excurrent opening, is generated by the beating of tiny hairs or cilia that line the pharynx and its perforations. Food particles, including phytoplankton and bacteria, are filtered as the water passes through the sieve-like pharynx. Eggs, sperm, and waste products are expelled from the excurrent opening. The excurrent openings of many colonial species empty into large internal spaces or canals instead of passing directly to the exterior.

Solitary ascidians have external fertilisation, the eggs and sperm being expelled into the surrounding waters through the excurrent siphon. Tadpole-like larvae develop in the plankton and are an important food source for damselfishes and other reef species. After only a few hours, they attach themselves to a substrate, absorb their tail, and develop into a juvenile sea squirt. Colonial species are internally fertilised. Embryos are incubated within the colony, either in the parent or in the matrix, and are expelled after they become tailed larvae. However, the free-swimming stage is very short-lived usually lasting less than one hour before metamorphosis takes place.

Opposite: Multicoloured sea squirts (blue *Rhopalaea sp.* and red *Didemnum sp.*) carpet Southeast Asian reefs. Photo by Gary Bell.
Below (top): The delicate Fourspot sea squirt (*Clavelina sp.*) is a 2-cm-high solitary form.
Bottom: This Yellow ascidian (unidentified styelid) is a colonial form composed of numerous individuals called zooids. It is growing on red sponge.

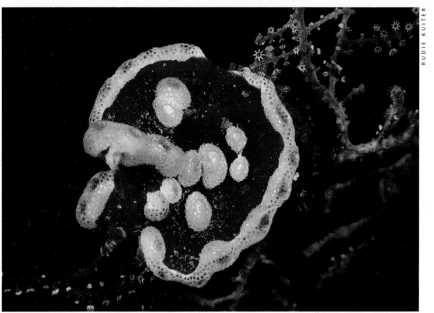

SEA SQUIRTS

1 Yellow Sea Squirt (8 cm)
Phallusia julinea; Ascidiidae

This common sea squirt is variable in shape, but can be recognised by the bright yellow colour and fringed lobes that surround the incurrent and excurrent openings. It occurs in a variety of habitats to depths of 35 metres but is most common in shallow water. Sometimes only the openings are evident, the rest of the body being firmly wedged and hidden among coral or rubble.

2 Bluebell Sea Squirt (colony width 12 cm)
Clavelina moluccensis; Clavelinidae

Colonies are usually seen attached to dead coral or rubble. Colour ranges from light to medium blue. A characteristic pattern of dark blue spots and patches is always present on the top of the zooids. Individual zooids vary between 0.5–2.5 cm in length. They are semi-transparent and often clustered in groups that are attached to a common base.

3 Robust Sea Squirt (colony width 6 cm)
Clavelina robusta; Clavelinidae

This species was given its scientific name only recently. Colonies are frequently large, with large (2–4 cm) robust zooids. Colour is generally black and individual zooids have very distinctive fluorescent green rings around the incurrent and excurrent openings. The outer covering is soft and slightly transparent but the basal matrix is firm.

4 Soft Sea Squirt (individual colonies 2 cm height)
Didemnum molle; Didemnidae

This is one of the most frequently noticed species of sea squirt occurring in a variety of reef habitat. The small (about 2 cm) spherical growths look like solitary individuals at first glance but each is a colony composed of tiny embedded zooea which have a common excurrent opening. The colonies may be widely separated or in grape-like clusters. Colour is generally green due to the presence of symbiotic algae.

5 Slimy Sea Squirt (colony width 12 cm)
Lissoclinum patella; Didemnidae

Some sea squirts have a very irregular shape and are difficult to recognise. For example, the Slimy sea squirt at first glance looks more like a sponge. The encrusting sheets are usually seen in lagoons or other semi-protected environments where they grow on coral rock. They sometimes surround living corals, eventually blocking out the light and killing them.

6 Transparent Sea Squirt (4 cm)
Rhopalaea sp.; Diazonidae

The transparent outer covering of this solitary sea squirt allows a glimpse of the intricate sieve-like apparatus that occupies much of the body cavity. This species is common on coral reefs at depths ranging from about 2–20 metres. Individuals are elongated, reaching a length of up to 6 cm.

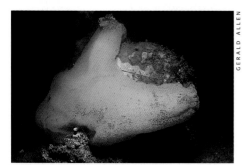

1 Yellow Sea Squirt *Phallusia julinea*

4 Soft Sea Squirt *Didemnum molle*

5 Slimy Sea Squirt *Lissoclinum patella*

2 Bluebell Sea Squirt *Clavelina moluccensis*

3 Robust Sea Squirt *Clavelina robusta*

6 Transparent Sea Squirt *Rhopalaea sp.*

7 Laysan Sea Squirt (colony width 12 cm)
Eudistoma laysani; Polycitoridae
This colonial sea squirt has the appearance of a soft coral from the distance. A number of fleshy stalks or lobes radiate from a central mass that is attached to the bottom. Each lobe contains up to 100 zooids but usually there are less than 50. When fully inflated the separate lobes appear to join, forming one continuous colony. The tips of the lobes, from which the numerous incurrent and excurrent openings protrude, are semi-transparent with a bluish sheen.

8 Tentacle Sea Squirt (colony width 10 cm)
Pycnoclavella detorta; Pycnoclavellidae
This is a colonial species with 3-cm-long zooids rising on stalks from a common base. The delicate tentacle-like zooids are partly transparent with greenish iridescent patches around the incurrent opening. The habitat consists of coastal reefs and outer slopes to depths of 40 metres.

9 Lightbulb Sea Squirt (part of colony, width 5 cm)
Pycnoclavella diminuta; Pycnoclavellidae
Colonies are formed of spherical, 5-mm-diameter zooids on short stalks that attach to a common base. The zooids are soft, semi-transparent and in a range of colours that include white, yellow, orange, blue, purple and blackish. It inhabits caves and ledges at depths between about 5–20 metres.

10 Tomato Sea Squirt (colony width 12 cm)
Eusynstyela sp. Styelidae
The favoured habitat of this bright red species are caves and crevices, where it forms encrusting growths amongst sponges, daisy corals and bryozoans. Individual zooids are 1–2 cm long and protrude from the encrusting mat. Each has a distinct pair of openings, which are surrounded by a more intense shade of red.

11 Leach's Sea Squirt (colony width 10 cm)
Botrylloides leachi; Styelidae
This species is among the most variable of the sea squirts with regards to colour pattern and there seems to be no limit to the variations. It is a colonial species forming delicate encrusting sheets. The tiny zooids are numerous and tightly packed within the matrix. They form long curving or branching double-row systems as well as circular or oval systems.

12 Goldmouth Sea Squirt (8 cm length)
Polycarpa aurata; Styelidae
This is a very common species throughout the region and probably the most commonly photographed sea squirt. Colour is extremely variable ranging from whitish to brilliant shades of blotchy yellow and purple. The firm outer coat is marked with purple grooves and the interior surface of the openings are yellow. It attaches itself to rocky surfaces or exposed, dead coral and under ledges between 10–20 metres.

7 Laysan Sea Squirt *Eudistoma laysani*

8 Tentacle Sea Squirt *Pycnoclavella detorta*

12 Goldmouth Sea Squirt *Polycarpa aurata*

9 Lightbulb Sea Squirt *Pycnoclavella diminuta*

10 Tomato Sea Squirt *Eusynstyela sp.*

11 Leach's Sea Squirt *Botrylloides leachi*

Fishes
A Kaleidoscope of Colour

Fishes are certainly the most conspicuous of all the reef's inhabitants and no other region on earth can match the seemingly endless variety found in the seas of Southeast Asia.

Thanks to a long history of biological exploration in the area and, more recently, the concentrated efforts of a small band of devoted scuba diving scientists and underwater photographers, reef fishes are better known than most of the other marine creatures of Southeast Asia. This section will assist with the identification of representatives from most of the common families. In addition, several comprehensive references are listed at the end of the book. The extra effort involved in learning the commonly observed species will infinitely enrich your diving holiday.

The reef fishes of Southeast Asia are an intimate part of the huge Indo-Pacific fish community. Nearly all the families, and genera and many individual species, are distributed throughout the tropical Indian and Pacific Oceans. This means there is a certain similarity in the fishes whether one is diving at Hawaii, Indonesia, the Maldives or in the Red Sea. Indeed, many individual species occurring in Indonesia, for example the Raccoon Butterflyfish (page 74), are found throughout this vast region. This is certainly a great advantage when learning the fishes of an area, as there is always a degree of familiarity at each new location, based on past experience elsewhere.

The observant diver will notice the same individual fishes, often restricted to a small section (a few square metres or less) of the bottom, if the same reef is visited over a period of successive days or even successive years.

How then, do these fishes, which are anchored to a

Opposite: Fearless cleaner shrimp (*Lysmata amboinensis*) enter the mouth of a grouper (*Cepholopholis miniata*) while searching for parasites. Although small crustaceans are normally eaten by groupers, special privileges are extended to these parasite removers. Photo by Rudie Kuiter.

In The Steps of Pieter Bleeker

The influence of Pieter Bleeker on our present day knowledge of the fishes of Southeastern Asia is profound. No single individual has done so much to further our knowledge of the region's fishes, nor is it likely that his exploits will ever be matched.

He arrived in Batavia (now Jakarta) in 1843, fresh out of medical school in Holland, as a 23-year-old recruit in the colonial army of the Dutch East India Company. Shortly after his arrival he began making small fish collections. What started as a hobby soon became a driving passion. Bleeker amassed a fantastic collection based on his own efforts and the help of other army personnel scattered throughout the archipelago. Considering his obligations as a full-time army surgeon, the extent of Bleeker's activities is absolutely amazing. He spent 17 years in Indonesia and continued his studies of Southeast Asian fishes after his return to Holland until his death in 1876. During a career spanning 36 years, Bleeker published some 500 scientific articles. Even more impressive is the total of 3,334 new species described in these published works. About three-fourths of these were from the Southeast Asian region.

It's easy to experience the same thrill of discovery that Bleeker must have felt in his early years on Java. All that is required is a mask and snorkel and a swim over any one of Southeast Asia's thousands of coral reefs. Yet, Bleeker appeared on the scene a century before modern diving gear made an appearance so it is really extraordinary to believe all his discoveries were made without ever having dived. Instead he relied on his network of contacts with army colleagues, fishermen and markets.

Although Bleeker encountered most of the common species of the region, many small or secretive fishes managed to elude him. Therefore, Southeast Asia has been a particularly fertile hunting ground for scuba diving scientists over the past two decades. Perhaps 200–300 additional new reef fishes have been captured during this period and many more await discovery. For example, a joint Indonesian-Australian expedition to the Indonesian island of Flores in November 1993 yielded seven species previously unknown to science.

Below: This anglerfish (*Antennarius biocellatus*) is one of hundreds of Indonesian fishes illustrated in Pieter Bleeker's monumental *Atlas Ichthyologie.*

small plot of reef become distributed to such far flung localities as Hawaii and East Africa? The answer to this riddle is found in the tiny larval stage of development. Most reef fishes, except for a small number of live-bearers, rely on the pelagic or oceanic larval stage for dispersal to other regions.

Studies that involve the counting of daily growth rings on the microscopic ear bones (the otoliths) of larval reef fishes indicate that individual species have an oceanic stage that lasts for up to several weeks. During this period, the transparent larvae, which are only a

few millimetres long, are at the mercy of ocean currents and may drift for hundreds of kilometres before colonising a shallow reef. Some fishes may actually prolong the length of their larval stage if a suitable shallow reef is not encountered.

The majority of reef fishes are egg layers that exhibit two main types of reproductive behaviour. However, both have a pelagic larval stage. The first group, typified by wrasses and parrotfishes, scatter large numbers of tiny, positively buoyant eggs into open water.

The second pattern involves species that lay eggs on the bottom, in rocky crevices, empty shells or other hard surfaces. Damselfishes, blennies, gobies and triggerfishes are among the best known examples. The parents (often only the male) care for the eggs until hatching. Male cardinalfishes and pipefishes (including seahorses), show an unusual variation on this theme, incubating the eggs in their mouth or on a special region on the underside of the body.

To the untrained observer, and especially a novice diver or snorkeller, the array of fishes encountered on a typical Southeast Asian coral reef is baffling and overwhelming. The amazing cavalcade of colours and shapes seems unbelievable, particularly if one's previous experience is confined to temperate latitudes.

What makes this region so special for fishes and other marine life is an incredible cumulative effect. Because no two reefs are ever exactly alike, they present different ecological opportunities, hence they may provide a home to quite different species. In addition, there

Below: Large schools of anthias (family Serranidae) are an integral part of Indo-Pacific fish communities.

ROGER STEENE

Coral Reef Fish Families

A brief profile of the 11 most common families on a typical reef is presented below.

Gobies (Family Gobiidae)
Small, bottom dwelling fishes particularly common in sand and rubble habitats. By far the largest family of marine fishes; over 300 species in Southeast Asia.

Wrasses (Family Labridae)
Colourful predators of crustaceans and other invertebrates, wrasses are common on reefs, in weed beds, and over sand bottoms. Female to male sex change is typical and each sex differs in colour. About 170 species in Southeast Asia.

Damselfishes (Family Pomacentridae)
Possibly the most numerous group as far as abundance of individuals is concerned, and therefore the most conspicuous. Includes the colourful clownfishes. Approximately 150 Southeast Asian species.

Cardinalfishes (Family Apogonidae)
The nocturnal equivalent of damselfishes as far as abundance is concerned. During the day they shelter in caves, crevices or around coral formations. About 100 species in Southeast Asia.

Groupers and anthias (Family Serranidae)
An incredibly diverse assemblage including both two-metre long groupers and the delicate anthias, some of

which only grow to 5 cm. Groupers are prominent on menus in the region's restaurants. Anthias are favourites of photographers and display an amazing harem-like social structure. Males are usually accompanied by a group of females but if anything should happen to the prominent male, the leading female changes quickly to the male sex. About 100 species in SoutheastAsia.

Surgeonfishes (Family Acanthuridae)
Often seen in schools, surgeonfishes graze on the filamentous algal mat that covers the reef. Avoid handling these fishes as they have a sharp scalpel-like spine on each side of the tail base. About 50 species in Southeast Asia.

Blennies (Family Blenniidae)
Similar to gobies in shape and habits, but lack scales. Common in tide pools and some species can even skip over the rocks from pool to pool. Both blennies and gobies lay eggs on the bottom and guard the nest for several days. About 75 species found in Southeast Asia.

Butterflyfishes (Family Chaetodontidae)
A favourite of divers and aquarists because of their beautiful shape and colours. Usually seen in pairs or small groups, which forage for live coral, algae and

is a tremendous wealth of non-reef habitats such as seagrass beds, rubble reefs, sand or mud flats, and brackish mangrove estuaries. Each of these environments has a largely unique assemblage of fishes, although they are not as species-rich as coral reefs. The extraordinary effect of these highly variable environments is largely responsible for Southeast Asia's colossal biodiversity.

A typical well-rated dive site in Southeast Asia, such as the famous shipwreck at Tulamben, Bali, is inhabited by 250–350 readily observed fish species. But due to the above mentioned cumulative effect there are perhaps 2,000 reef species scattered throughout the region. In spite of this impressive total, it is well within the grasp of the average diver to learn most of the common reef species.

The best approach is to first focus on the common families. Surprisingly, most of the reef's conspicuous fishes belong to a relatively small number of families. Families are best recognised on the basis of shape and behaviour. The butterflyfishes are probably the easiest of all groups to identify. Their rounded and laterally compressed shape, prominent dorsal spines, colourful patterns and graceful swimming behaviour are very distinctive.

More species of reef fishes have been found in the vicinity of Maumere Bay on the Indonesian island of Flores than any other place on earth. In spite of recent earthquake and tidal wave damage, it is still a great dive site.

Scientists have been monitoring Maumere's fish populations over the past eight years. An effort was made to survey every reef habitat and adjacent environments such as sand and mud flats, seagrass beds and the coastal mangroves. The results are very impressive with over 1,336 species recorded so far. This work has also resulted in the collection of approximately 20 fishes previously unknown to science. It is anyone's guess as to how many additional ichthyological treasures await discovery in the vast and species-rich Southeast Asian region.

Below: Jawfishes (Opistognathidae) are among the few marine families that incubate eggs in their mouth.

crustaceans. An estimated 60 species found in Southeast Asia.

Snappers (Family Lutjanidae)
Voracious predators of small fishes and invertebrates. They are excellent food fishes, commonly seen both on reefs and in the region's fish markets. About 50 species found in Southeast Asia.

Pipefishes (Family Syngnathidae)
Small inconspicuous, non fish-like creatures that swim close to the bottom or inhabit crevices. The group includes the popular seahorses. Males incubate eggs in a pouch, or special region on the belly or tail. About 50 species found in Southeast Asia.

Parrotfishes (Family Scaridae)
Closely related to wrasses, but most have larger scales and the teeth fused into a curious beak-like apparatus. They feed on filamentous algae and in this process consume huge amounts of coral rock that is pulverised by special teeth in the throat. The resultant sand is voided with the faeces and makes a huge contribution to the overall bottom sediment. About 45 Southeast Asian species.

Percentages of Major Fish Groups

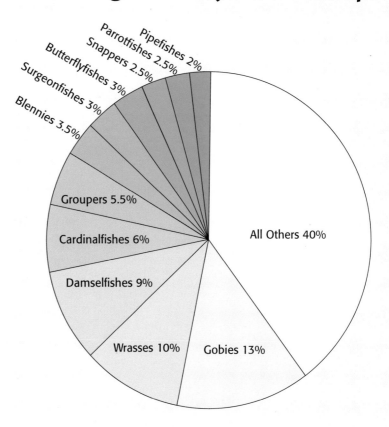

Pipefishes 2%
Parrotfishes 2.5%
Snappers 2.5%
Butterflyfishes 3%
Surgeonfishes 3%
Blennies 3.5%
Groupers 5.5%
Cardinalfishes 6%
Damselfishes 9%
Wrasses 10%
Gobies 13%
All Others 40%

1 Blackspot Squirrelfish (20 cm)
Sargocentron cornutum; Holocentridae
Squirrelfishes are nocturnal hunters of small fishes and crustaceans. During the day they are seen in caves, under ledges or sheltering near the bottom. They are characterised by a large eye, red to silvery colouration, and very rough scales. In addition they are capable of producing clearly audible sounds. These fishes should be handled with care as the sharp spines on the fins and head can cause painful wounds. Southeast Asian reefs are inhabited by about 30 species.

2 Ribbon Eel, adult female (100 cm)
Rhinomuraena quaesita; Muraenidae
One of the most unusual morays encountered, it lives in a sandy burrow, projecting the head and anterior part of the body out of its hole. In spite of the small head it has a menacing appearance with enlarged tentacle-like nostrils. The black and yellow males and juveniles were once thought to be a different species.

3 Spotted Garden Eel (40 cm)
Heteroconger hassi; Heterocongridae
Several species of garden eels inhabit sandy slopes next to coral reefs, often in huge colonies that include hundreds of individuals. Members of the colony live in burrows that are in close proximity. They never leave their sanctuary but stretch above the bottom when feeding on microscopic organisms conveyed by the currents.

4 Striped (Striped Eel) Catfish, juveniles (8 cm)
Plotosus lineatus; Plotosidae
Most catfishes in Southeast Asia are confined to fresh or brackish water, but the species shown here is a common reef inhabitant. It is easily recognised on the basis of its barbels or feelers that surround the mouth. Large aggregations of juveniles are often seen busily feeding in vacuum cleaner fashion. Avoid handling these fish as they have venomous fin spines.

5 Reef Lizardfish (14 cm)
Synodus variegatus; Synodontidae
Several types of lizardfishes are common on reefs and adjacent rubble. They are lie-and-wait predators that feed mainly on small fishes. However, they are able to launch themselves at amazing speed and seize unsuspecting prey.

6 Giant Moray (250 cm)
Gymnothorax javanicus; Muraenidae
Moray eels are common on all reefs in the region. They are specialised creatures, differing from most other fishes in several important respects. The snake-like body, lack of scales, absence of pectoral fins, and single gill opening behind the head are typical features. Relatively few morays are seen by divers. Most of the 60-odd species in Southeast Asia are shy, secretive dwellers of crevices and caverns. The Giant moray is the most commonly encountered and largest species.

1 Blackspot Squirrelfish *Sargocentron cornutum*

2 Ribbon Eel, adult female *Rhinomuraena quaesita*

3 Spotted Garden Eel *Heteroconger hassi*

4 Striped Catfish, juveniles *Plotosus lineatus*

6 Giant Moray *Gymnothorax javanicus*

5 Reef Lizardfish *Synodus variegatus*

7 Razorfish or **Shrimpfish** (10 cm)
Aeoliscus strigatus; Centriscidae

This bizarre creature bears little resemblance to a typical fish. It has a hard exterior surface and small transparent fins. Even more unusual, it spends most of the time "standing" on its head. It is typically found in aggregations, which move along the reef in a synchronised manner. They swim in a conventional horizontal mode only when threatened but quickly assume the vertical stance after fleeing to safety. The tiny mouth is adapted for feeding on shrimps and other crustaceans.

8 Smooth Flutemouth (100 cm)
Fistularia petimba; Fistulariidae

This fish is identified by its tubular snout, elongate and cylindrical body, and the trailing filament on the tail. It relies on its nondescript colouration and a very slow, unorthodox, swimming motion to sneak up on unsuspecting prey, usually juvenile fishes.

9 Trumpetfish (40 cm)
Aulostomus chinensis; Aulostomidae

The Trumpetfish is a relative of the flutemouth and has a similar snout and elongate shape. However, it lacks the long tail filament. The colour is variable, ranging from a mottled greenish brown to uniform bright yellow. It often swims side by side with other large fish such as parrotfishes or goatfishes. This strategy is used to approach within striking range of its fish prey.

10 Smallscale Scorpionfish (20 cm)
Scorpaenopsis oxycephalus; Scorpaenidae

This species also has toxic fin spines, similar in structure to those of *Pterois*. The poison is produced by skin cells that cover grooves running the length of each spine. The protein-based venom is unstable at high temperatures and is broken down, therefore it is important to administer the hot water-soaking treatment to sting victims. The Smallscale scorpionfish is a master of disguise. Lying motionless while waiting for fish prey, it is nearly indistinguishable from its surroundings.

11 Leaf Scorpionfish (8 cm)
Taenianotus triacanthus; Scorpaenidae

This fish is relatively common, but easily overlooked due to its excellent camouflage colours and its habit of swaying from side to side with the rhythm of the current. Colouration is highly variable.

12 Red Firefish or **Lionfish** (28 cm)
Pterois volitans; Scorpaenidae

This spectacular fish is commonly encountered on most reefs. It can be approached at close range, but beware of its highly venomous spines. All of the fin spines are poisonous and the puncture wounds are extremely painful. Victims may lapse into a coma. Fortunately, the venom can be effectively neutralised by immediate soaking of the wound in very hot water until the pain subsides.

7 Razorfish or **Shrimpfish** *Aeoliscus strigatus*

10 Smallscale Scorpionfish *Scorpaenopsis oxycephalus*

11 Leaf Scorpionfish *Taenianotus triacanthus*

8 Smooth Flutemouth *Fistularia petimba*

9 Trumpetfish *Aulostomus chinensis*

12 Red Firefish or **Lionfish** *Pterois volitans*

FISHES

13 Coral Grouper (30 cm)
Cephalopholis miniata; Serranidae
The fishes appearing on this page are members of the family Serranidae and are abundant on reefs throughout the region. The group is readily divisible into two major subfamilies: the coral trouts and groupers belonging to the subfamily Epinepheliniae and the anthias or fairy basslets of the subfamily Anthiinae. The Coral grouper is a common inhabitant of caves and ledges and preys on small fishes and crustaceans.

14 Giant or **Queensland Grouper** (200 cm)
Epinephelus lanceolatus; Serranidae
Most of the groupers range between about 25–40 cm when fully grown, but the Giant grouper may reach a length of 2.5 metres and a weight of 400 kg. This fish is an unforgettable sight. It has the unnerving habit of slinking up to an unsuspecting diver whose back is turned. To wheel around and face this monster at close range is guaranteed to get the adrenaline pumping!

15 Coronation Grouper (50 cm)
Variola louti; Serranidae
The Coronation grouper and its related species are much in demand by fish markets and restaurants. It is now reared commercially in SE Asia to ensure adequate supplies. Farms consist of a floating bamboo framework from which nets are suspended to form compartments or pens.

16 Redfin Anthias, male (7 cm)
Pseudanthias ignitus; Serranidae
The delicate anthias form large colourful schools that swim high above the bottom. Unlike their fish-eating grouper cousins, their diet consists mainly of tiny zooplankton conveyed by the currents. They exhibit an unusual harem-type social structure whereby a single dominant male reigns supreme over a group of females. The Redfin anthias is common on the upper edge of outer reef slopes.

17 Squarespot Anthias, male (12 cm)
Pseudanthias pleurotaenia; Serranidae
A hierarchy exists among the females of an anthias' harem. The largest and most aggressive fish is subservient only to the male. These harems are readily apparent to divers as the male and female show significant colour pattern differences. If the male is experimentally removed from the group or eaten by a predator, the dominant female changes to the male sex and assumes its colouration over a period of several days. Consequently, each member of the harem then climbs another step up the pecking order.

18 Squarespot Anthias, female (10 cm)
Pseudanthias pleurotaenia; Serranidae
This species inhabits steep outer reef slopes, usually below depths of 15–20 metres. It feeds in midwater but quickly retreats to rocky crevices if disturbed.

13 Coral Grouper *Cephalopholis miniata*

14 Giant Grouper *Epinephelus lanceolatus*

15 Coronation Grouper *Variola louti*

18 Squarespot Anthias *Pseudanthias pleurotaenia*

17 Squarespot Anthias *Pseudanthias pleurotaenia*

16 Redfin Anthias *Pseudanthias ignitus*

19 Royal Dottyback (6 cm)

Pseudochromis paccagnellae; Pseudochromidae

The dottybacks include some of the most beautiful of all coral reef fishes, but due to their small size (usually less than 8 cm) and secretive way of life, they are seldom noticed. There are approximately 30 species in the Southeast Asian region. These fishes always stay close to shelter, usually rocky crevices, although they sometimes seek refuge in hollow sponges. Several new species have recently been discovered in the region.

20 Ring-tailed Cardinalfish (10 cm)

Apogon aureus; Apogonidae

The cardinalfishes are one of the largest families of coral reef fishes. Nearly all of the species seen on coral reefs are nocturnally active. During daytime, they shelter close to the reef's surface, but at night are seen swimming out in the open, high above the bottom. They feed primarily on a variety of small crustaceans and also copepods, common micro-crustaceans that form part of the plankton.

21 Coral Cardinalfish (6 cm)

Sphaeramia nematoptera; Apogonidae

During the day the Coral cardinalfish forms resting aggregations close to the bottom in rich coral areas. All cardinalfishes exhibit the unusual habit of oral egg incubation. The female releases a ball containing a few hundred to several thousand eggs. These are picked up by the male who then keeps them in his mouth for several days until hatching occurs.

22 Bigeye Trevally (70 cm)

Caranx sexfasciatus; Carangidae

The trevallies or jacks belong to a large family of silvery, fast swimming fishes distributed worldwide on both temperate and tropical reefs. They are highly esteemed as food fishes. There are at least 60–70 species found in this region. Only a small portion of these are commonly encountered on coral reefs. The Bigeye trevally is perhaps the most conspicuous because of its habit of forming large schools that approach divers at close range.

23 Blue Blanquillo (30 cm)

Malacanthus latovittatus; Malacanthidae

The malacanthids or tilefishes are a small family of elongate reef fishes that prefer sand or rubble bottoms. They construct burrows and frequently occur in pairs. The juvenile of this species has a broad black stripe from the tip of the snout to the tail fin.

24 Bluefin Trevally (40 cm)

Caranx melampygus; Carangidae

The common name of this species is derived from its electric blue fins. It is frequently seen on coral reefs, most often solitarily or in small groups. They are swift predators of other fishes. Trevallies feed most actively at dawn and dusk. Prey are disadvantaged at these times

19 Royal Dottyback *Pseudochromis paccagnellae*

22 Bigeye Trevally *Caranx sexfasciatus*

23 Blue Blanquillo *Malacanthus latovittatus*

20 Ring-tailed Cardinalfish *Apogon aureus*

21 Coral Cardinalfish *Sphaeramia nematoptera*

24 Bluefin Trevally *Caranx melampygus*

FISHES

due to lower light levels and the confusion of the change over when nocturnal fishes retire and daytime species emerge from their hiding places.

25 Paddletail Snapper (35 cm)
Lutjanus gibbus; Lutjanidae
Snappers are common on reefs throughout the region. They are distinguished from similar families by having a scaly cheek, 10 dorsal spines, and most species have distinct canine teeth. The Paddletail snapper is easily recognised by its forked tail fin which has rounded lobes. It is most frequently seen in clear water on outer reef slopes. This species is sometimes implicated in cases of human fish poisoning known as ciguatera.

26 Yellowstripe Snapper (30 cm)
Lutjanus kasmira; Lutjanidae
This distinctive species is easily recognised. It is common on most reefs, sometimes forming stationary schools during the day on shipwrecks or around coral formations. The species ranges widely in the Indo-Pacific region from East Africa to Polynesia.

27 Midnight Snapper (35 cm)
Macolor macularis; Lutjanidae
The Midnight snapper and closely related Black and White snapper (*Macolor niger*) are common on the upper edge of drop-offs. Juveniles and subadults are distinguished by a black upper half and white lower half.

The small young, which swim with a peculiar bobbing and swaying motion, are usually found near Gorgonian sea fans or Black coral.

28 Sailfin Snapper (30 cm)
Symphorichthys spilurus; Lutjanidae
This spectacular fish differs from most other snappers by virtue of its taller body and very elongate dorsal fin rays. It also has a very pronounced false eye spot or ocellus at the base of the tail. It is most often seen in sand or rubble areas near reefs.

29 Many-spotted Sweetlips (30 cm)
Plectorhinchus chaetodontoides; Haemulidae
The sweetlip family is represented on SE Asian coral reefs by about one dozen species. These snapper-like fishes are distinguished by their very thick lips. Most are solitary fishes but a few species form large schools. It undergoes a dramatic change from juvenile to adult.

30 Blue and Gold Fusilier (30 cm)
Caesio teres; Caesionidae
Fusiliers are closely related to snappers. However, they are more streamlined in shape and usually swim in open water well above the bottom. They form big, colourful schools sometimes composed of several species and frequently swarm around divers. Most are vivid shades of blue and there is usually a yellow patch on the back or one or more yellow stripes on the side of the body.

25 Paddletail Snapper *Lutjanus gibbus*

26 Yellowstripe Snapper *Lutjanus kasmira*

27 Midnight Snapper *Macolor macularis*

30 Blue and Gold Fusilier *Caesio teres*

28 Sailfin Snapper *Symphorichthys spilurus*

29 Many-spotted Sweetlips *Plectorhinchus chaetodontoides*

31 Teira Batfish (30 cm)
Platax teira; Ephippidae
Juveniles of this species have incredibly tall dorsal and anal fins, but adults (shown here) are more round. The adults are often seen in schools on outer reef slopes or around shipwrecks. Batfishes have very peculiar brush-like teeth with tricuspid tips. They feed mainly on zoo-plankton and benthic or bottom-dwelling invertebrates.

32 Big-eye Bream (30 cm)
Monotaxis grandoculis; Lethrinidae
The members of this family are commonly known as emperors. They are conspicuous inhabitants of the sandy fringe of coral reefs. The general appearance is similar to that of snappers. Some species, such as the big-eye bream, have molar-like teeth on the side of the jaws. These are useful for crushing sand-dwelling invertebrates including molluscs, worms, crabs and shrimps. Juveniles of this species are largely white with three broad, black saddles on the back.

33 Two-lined Monocle Bream (20 cm)
Scolopsis bilineatus; Nemipteridae
The monocle bream and their relatives of the family Nemipteridae are often seen around the sandy edge of coral reefs. They are very active fishes that continually search the bottom for small invertebrate food. Their locomotion consists of quick darting movements inter-spersed with brief periods of stationary hovering.

34 Yellow-lined Goatfish (25 cm)
Mulloidichthys vanicolensis; Mullidae
Goatfishes are easily distinguished by the pair of feelers or barbels on the chin, the elongate body and two sepa-rate dorsal fins. They occur on reefs and adjacent sand or rubble bottoms. The barbels have taste buds and are used to probe the bottom for small invertebrate food. About 20 species of goatfishes are found on Southeast Asian reefs.

35 Topsail Drummer (30 cm)
Kyphosus cinerascens; Kyphosidae
This family, popularly known as drummers or rudder-fishes, contains only 2–3 common species in Southeast Asia and the Pacific region. They occur both individu-ally or in schools, usually close to shore. Although they are omnivores, the main food consists of algae. Their digestive tract is very long: a typical feature of plant-eating fishes.

36 Round-faced Batfish (10 cm)
Platax orbicularis; Ephippidae
Batfishes are one of the reef's genuine personality fishes. Whether alone or in schools, they never fail to impress the onlooker. They are frequently curious and will approach closely, sometimes even accepting food from the hand of divers. Juveniles (which are shown here) frequently occur around boat moorings or off sandy beaches.

31 Teira Batfish *Platax teira*

32 Big-eye Bream *Monotaxis grandoculis*

33 Two-lined Monocle Bream *Scolopsis bilineatus*

36 Round-faced Batfish *Platax orbicularis*

35 Topsail Drummer *Kyphosus cinerascens*

34 Yellow-lined Goatfish *Mulloidichthys vanicolensis*

37 Threadfin Butterflyfish (20 cm)
Chaetodon auriga; Chaetodontidae

This fish is usually seen solitarily or in pairs, occasionally in small schools, on coral reefs or adjacent weed bottoms. It is popular in the aquarium trade and does well in captivity. The natural diet consists of algae and small invertebrates, particularly worms and shrimps. The species is very widely distributed in the Indo-Pacific region, from East Africa and the Red Sea to Polynesia.

38 Pacific Triangular Butterflyfish (14 cm)
Chaetodon baronessa; Chaetodontidae

This is a strongly territorial species that is usually seen alone or in pairs. It is invariably associated with corals belonging to the genus *Acropora*, which is also its favourite food. This species is common throughout most of Southeast Asia. However, it is replaced by a nearly identical species, the Indian triangular butterflyfish, *Chaetodon triangulum*, on the Indian Ocean coast of Thailand, Malaysia and Sumatra.

39 Bennett's Butterflyfish (16 cm)
Chaetodon bennetti; Chaetodontidae

Many butterflyfishes possess a false eye spot or ocellus. Presumably it is beneficial in directing potentially fatal attacks of predators away from the head to a less vulnerable body part. Also, this distinctive mark may frighten enemies away. Bennett's butterflyfish is frequently seen on outer slopes and dropoffs, usually in pairs.

40 Raccoon Butterflyfish (18 cm)
Chaetodon lunula; Chaetodontidae

Usually seen in pairs or small groups in rich coral areas. It is considered a home-ranging species that forages within a relatively wide area of a particular reef complex. Its diet largely consists of nudibranchs, tubeworms, algae, coral polyps and a variety of benthic invertebrates. The larval stage is characterised by a series of peculiar bony plates covering the head. This species is widely distributed from East Africa to Polynesia.

41 Ornate Butterflyfish (16 cm)
Chaetodon ornatissimus; Chaetodontidae

This is one of two similarly patterned species found in the region. The other, *Chaetodon meyeri*, is distinguished by black, rather than orange bands on the side and usually has a bluish cast to the body. Both species are often seen in pairs and occur in rich coral areas. They feed mainly on polyps. The distribution extends from East Africa to the Central Pacific.

42 Saddled Butterflyfish (20 cm)
Chaetodon ephippium; Chaetodontidae

The huge eyespot on the rear half of the body clearly identifies this species. It is usually sighted alone or in pairs. Scientific studies reveal these pairs frequently last for a lifetime (estimated between 8–15 years). Preferred food items include algae, sponges, coral polyps and small crustaceans.

37 Threadfin Butterflyfish *Chaetodon auriga*

40 Raccoon Butterflyfish *Chaetodon lunula*

41 Ornate Butterflyfish *Chaetodon ornatissimus*

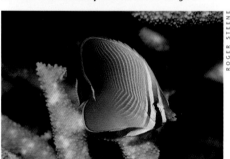

38 Pacific Triangular Butterflyfish *Chaetodon baronessa*

39 Bennett's Butterflyfish *Chaetodon bennetti*

42 Saddled Butterflyfish *Chaetodon ephippium*

43 Pacific Redfin Butterflyfish (12 cm)
Chaetodon lunulatus; Chaetodontidae
The habitat usually consists of rich coral gardens which provide its exclusive polyp food source. It is a butterflyfish that forms lifelong mated pairs. A similar species, the Indian redfin butterflyfish, *Chaetodon trifasciatus*, occurs off western Thailand and Malaysia, and at some Indonesian Islands, including Sumatra, Java and Bali. Both species are found at some Indonesian localities.

44 Chevroned Butterflyfish (18 cm)
Chaetodon trifascialis; Chaetodontidae
This is the most territorial of all butterflyfishes. Individuals establish a small territory, usually centred on a large *Acropora* table coral or several adjacent coral heads. It vigorously defends its turf, particularly against members of the same species. It feeds exclusively on the live polyps and mucus of staghorn and tabletop *Acropora* coral. It constantly grazes the coral, but does not kill it, thus insuring a continuous food supply.

45 Pacific Double-saddle Butterflyfish (14 cm)
Chaetodon ulietensis; Chaetodontidae
The twin dark bars on the upper part of the back serve to recognise this species. It is commonly sighted in pairs or small groups in rich coral areas or on outer reef drop-offs. It is a home-ranging species that forages over a limited territory for bottom-living invertebrates.

46 Longnose Butterflyfish (18 cm)
Forcipiger flavissimus; Chaetodontidae
This fish is commonly seen alone or in pairs on outer reef slopes. The distinctive elongate snout is used to probe reef crevices for hydroids, crustaceans and worms. A nearly identical fish, *Forcipiger longirostris*, has a slightly shorter snout. Both species have very broad geographic distributions.

47 Longfin Bannerfish (25 cm)
Heniochus acuminatus; Chaetodontidae
The Longfin bannerfish is usually encountered solitary or in small groups near the bottom. It has a distinctive dorsal fin that is greatly prolonged to form a trailing filament. A very similar species, *Heniochus diphreutes*, is best distinguished on the basis of behaviour. In contrast to *Heniochus acuminatus*, it forms schools that swim well above the bottom. Both species feed on a combination of zooplankton and small bottom-living invertebrates such as worms and shrimps.

48 Vagabond Butterflyfish (15 cm)
Chaetodon vagabundus; Chaetodontidae
The Vagabond is one of the most common butterflyfishes throughout the region. It forms permanent mated pairs that roam over sections of the reef in search of food. The diet includes algae, anemones, worms, coral polyps, sponges and tiny crustaceans. A very similar species, the Indian vagabond, *Chaetodon decussatus*,

43 Pacific Redfin Butterflyfish *Chaetodon lunulatus*

44 Chevroned Butterflyfish *Chaetodon trifascialis*

46 Longnose Butterflyfish *Forcipiger flavissimus*

45 Pacific Double-saddle Butterflyfish *Chaetodon ulietensis*

47 Longfin Bannerfish *Heniochus acuminatus*

48 Vagabond Butterflyfish *Chaetodon vagabundus*

The majestic **Blue-ringed angelfish** *(Pomacanthus annularis)* feeds mainly on sponges and algae. It ranges throughout Southeast Asia and adjacent seas.

The **Yellowkeel surgeonfish** *(Naso lituratus)* is a common and conspicuous reef inhabitant. At night it shelters in reef crevices.

inhabits the Indian Ocean coast of Thailand, Malaysia and Sumatra, and parts of southern Indonesia.

49 Pennant Bannerfish (12 cm)
Heniochus chrysostomus; Chaetodontidae
This attractive species has a similar shape to the Longfin bannerfish, but the dorsal fin is considerably shorter and broad at the tip. It is a shy species, usually seen sheltering close to the bottom, frequently under tabletop *Acropora* coral. It feeds mainly on coral polyps.

50 Bicolor Angelfish (8 cm)
Centropyge bicolor; Pomacanthidae
The angelfishes are close relatives of the butterflyfishes and were once considered to be in the same family. They differ from butterflyfishes in possessing a sharp spine on the lower edge of the cheek. Because of their small size compared to other members of the family—they are usually less than 10 cm—members of the genus *Centropyge* are commonly known as pygmy angels.

51 Yellowmask Angelfish (35 cm)
Pomacanthus xanthometopon; Pomacanthidae
Arguably the most attractive and spectacular of all the angelfishes, the Yellowmask is a shy, retiring species sometimes sighted on steep drop-offs. It always remains close to cover, usually a rocky overhang or cave. Its diet consists mainly of sponges. The geographic range extends from the Maldives to Melanesia.

52 Six-banded Angelfish (40 cm)
Pomacanthus sexstriatus; Pomacanthidae
This is the largest angelfish in the region, growing to about 46 cm. It is relatively common in many areas, frequenting both coastal and outer reefs, but most abundant in silty habitats. Divers are sometimes startled by the loud thumping staccato produced by the adults of this species and other large angelfishes. It is apparently a warning that is used to scare predators. The sound is so powerful that you can actually feel the vibrations.

53 Blue-girdled Angelfish (28 cm)
Pomacanthus navarchus; Pomacanthidae
This handsome angelfish is seen in lagoon and outer reef habitats throughout the Southeast Asian region. The orange body, broad blue bar on the head and on the front of the body are unmistakable features useful for identification. It feeds mainly on sponges supplemented with a variety of small bottom-living invertebrates.

54 Emperor Angelfish (35 cm)
Pomacanthus imperator; Pomacanthidae
This species, and all other large angelfishes, undergoes a dramatic transformation in colour pattern from the juvenile to adult stages. The young are blue-black with a stunning pattern of white concentric rings. This pattern gradually dissolves and the adult pattern shown here is assumed when a size of about 15–20 cm is reached.

49 Pennant Bannerfish *Heniochus chrysostomus*

50 Bicolor Angelfish *Centropyge bicolor*

51 Yellowmask Angelfish *Pomacanthus xanthometopon*

54 Emperor Angelfish *Pomacanthus imperator*

52 Six-banded Angelfish *Pomacanthus sexstriatus*

53 Blue-girdled Angelfish *Pomacanthus navarchus*

55 Regal Angelfish (23 cm)
Pygoplites diacanthus; Pomacanthidae

The Regal angelfish inhabits rich coral areas. Like other members of the family it prefers to stay close to shelter. It eats mainly sponges and sea squirts. The young are bright orange with a brilliant eye spot or ocellus on the posterior part of the dorsal fin. It is widely distributed from the Red Sea to the Central Pacific.

56 Indo-Pacific Sergeant (14 cm)
Abudefduf vaigiensis; Pomacentridae

Damselfishes are the third largest family of coral reef fishes and among the reef's most conspicuous inhabitants. Members of the genus *Abudefduf* are relatively large damselfishes commonly seen in a variety of reef habitats. The Indo-Pacific sergeant is often abundant on the upper edge of steep drop-offs or around shipwrecks.

57 Skunk Anemonefish (8 cm)
Amphiprion akallopisos; Pomacentridae

This species is primarily distributed in the Indian Ocean, ranging as far west as the African coast. In our region it is commonly encountered on the western coast of Thailand and Malaysia and also on the Indian Ocean side of Sumatra. It is also present at Java and Bali. It associates with only one anemone, *Heteractis magnifica*. Several individuals, including an adult pair, are usually found with each host. The depth distribution ranges from about 3–25 metres.

58 False Clown Anemonefish (8 cm)
Amphiprion ocellaris; Pomacentridae

This is the most popular species of anemonefish due to its striking colour pattern. It is commonly seen with two anemones, *Heteractis magnifica* and *Stichodactyla gigantea*. As in other anemonefishes, spawning occurs at regular, often monthly intervals. Nesting pairs are easy to detect because of their aggressive nature. They produce clicking sounds and will not hesitate to attack a diver.

59 Clark's Anemonefish (10 cm)
Amphiprion clarkii; Pomacentridae

Clark's anemonefish is the most widely distributed of all anemonefishes, ranging from the Persian Gulf to islands of the Western Pacific. In Southeast Asia it is most commonly seen with *Heteractis crispa* and *Stichodactyla mertensii*. Normally each anemone is inhabited by an adult pair and several small fish which are not necessarily juveniles. Studies reveal that the adult pair somehow inhibit the growth of the smaller fish. Anemonefishes are able to change from the male to female sex. There is a pecking order hierarchy among the smaller fish based on size. If the adult pair is experimentally removed, the gonads of the two largest fish become mature and they take over the reproductive role. Similarly, if only the female is removed the male changes sex and the next largest fish assumes the role of the male partner.

55 Regal Angelfish *Pygoplites diacanthus*

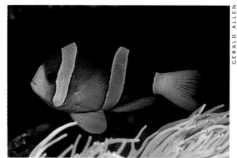
59 Clark's Anemonefish *Amphiprion clarkii*

60 Tomato Anemonefish *Amphiprion frenatus*

56 Indo-Pacific Sergeant *Abudefduf vaigiensis*

57 Skunk Anemonefish *Amphiprion akallopisos*

58 False Clown Anemonefish *Amphiprion ocellaris*

60 Tomato Anemonefish (8 cm)
Amphiprion frenatus; Pomacentridae
The Tomato anemonefish is commonly seen in the northern part of the region. It is usually found in company with the anemone *Entacmaea quadricolor.* The female is nearly twice the size of its brilliant red partner. The species ranges north to Japan.

61 Spinecheek Anemonefish (12 cm)
Premnas biaculeatus; Pomacentridae
This species is found exclusively with the anemone *Entacmaea quadricolor.* The female is usually at least twice the size of her male partner but duller. However, the colours quickly intensify if the fish is agitated.

62 Pink Anemonefish (8 cm)
Amphiprion perideraion; Pomacentridae
The sexes of most anemonefishes are very difficult to differentiate on the basis of colour. However, the males of the Pink anemonefish tend to have narrow orange margins on the rear part of the dorsal fin and upper and lower edges of the tail fin. This species is found exclusively with *Heteractis magnifica.*

63 Blue-green Chromis (6 cm)
Chromis viridis; Pomacentridae
Damselfishes of the genus *Chromis* are abundant throughout the region forming large aggregations that feed on zooplankton high above the bottom, or form dense clouds around coral heads. Males and females of the Blue-green chromis are similar in colour except during courtship and spawning. Sexually aroused males have a blackened dorsal fin and also the upper edge of the pectoral or breast fins becomes dark.

64 Humbug (or Domino) Dascyllus (4 cm)
Dascyllus aruanus; Pomacentridae
This distinctively marked damselfish is usually seen in aggregations that are closely associated with various types of branching corals. As many as 20–30 fish can be found in a single small coral head. It is most commonly seen in lagoons and on protected coastal reefs at depths between 1–12 metres.

65 Threespot Dascyllus (6 cm)
Dascyllus trimaculatus; Pomacentridae
The young of this fish are sometimes associated with anemones in the same manner as *Amphiprion* and *Premnas.* Their dependence on the anemone gradually decreases as the fish grow larger. Courtship in this species, as in other damselfishes, is characterised by rapid up and down swimming by the male. This behaviour, known as signal jumping, attracts females.

66 Blue Devil, male (8 cm)
Chrysiptera cyanea; Pomacentridae
This is one of a few damselfishes in which the sexes can be easily determined. Males usually have a bright yel-

61 Spinecheek Anemonefish *Premnas biaculeatus*

64 Humbug Dascyllus *Dascyllus aruanus*

65 Threespot Dascyllus *Dascyllus trimaculatus*

62 Pink Anemonefish *Amphiprion perideraion*

63 Blue-green Chromis *Chromis viridis*

66 Blue Devil, male *Chrysiptera cyanea*

low tail and lack a small black spot found at the base of the dorsal fin in females. It prefers depths less than 10 metres on protected coastal reefs and in lagoons. Nesting usually takes place in rock holes or crevices.

67 Neon Damsel (5 cm)
Pomacentrus coelestis; Pomacentridae
Especially attractive in large aggregations, the vivid blue Neon damsel is very common on rubble bottoms and in areas exposed to moderate wave action that supply planktonic food. The Goldbelly damsel, *Pomacentrus auriventris*, is a similar species occurring in the same habitat. It is easily distinguished by its bright yellow colouration on the lower sides and tail.

68 Lemon Damsel (5 cm)
Pomacentrus moluccensis; Pomacentridae
This fish is extremely abundant in rich coral areas, both close to shore and on outer slopes. It feeds on planktonic organisms a short distance above the bottom, but quickly retreats among the coral branches when threatened. The distribution encompasses a wide area in the eastern Indian Ocean and Western Pacific.

69 Blackside Hawkfish (15 cm)
Paracirrhites forsteri; Cirrhitidae
About a dozen species of hawkfishes are found on reefs in the region. The family is related to the groupers. They are lie-and-wait predators of juvenile fishes and

small crustaceans, typically seen sitting motionless on rocky outcrops for long periods.

70 Axilspot Hogfish (18 cm)
Bodianus axillaris; Labridae
The wrasse family is the second largest in the region. Representatives are encountered in all reef habitats, including surrounding sand and rubble bottoms or weed and seagrass beds. Hogfishes belonging to the genus *Bodianus* are one of the most conspicuous types of wrasses encountered by divers.

71 Redbreasted Wrasse (28 cm)
Cheilinus fasciatus; Labridae
The Redbreasted wrasse is frequently seen on outer reefs. It belongs to the genus *Cheilinus*, sometimes referred to as Maori wrasses for their striking facial patterns. There are approximately 15 species representing this genus in our region.

72 Chevron Barracuda (100 cm)
Sphyraena qenie; Sphyraenidae
Several types of barracudas are found in the region, but this is one of the most impressive due to its habit of forming spectacular aggregations next to outer reef slopes. Large barracudas have been implicated in human attacks in the Caribbean, but no attacks are documented in Southeast Asia. They have razor-sharp teeth and feed primarily on fishes.

67 Neon Damsel *Pomacentrus coelestis*

70 Axilspot Hogfish *Bodianus axillaris*

71 Redbreasted Wrasse *Cheilinus fasciatus*

68 Lemon Damsel *Pomacentrus moluccensis*

69 Blackside Hawkfish *Paracirrhites forsteri*

72 Chevron Barracuda *Sphyraena qenie*

73 Humphead or **Giant Wrasse** (190 cm)
Cheilinus undulatus; Labridae
This is truly the king of the wrasse family, growing to the huge size of 230 cm and approximately 200 kg. It is sometimes confused with the Bumphead parrotfish, but the snout is more pointed and lacks the fused beak-like teeth of the parrotfish. In spite of its huge size it is a very shy fish that is difficult to approach. It feeds on a variety of molluscs, fishes, sea urchins and crustaceans, and is one of the few predators of the notorious Crown-of-thorns starfish.

74 Yellowtail Coris, juvenile (4 cm)
Coris gaimard; Labridae
The juvenile of this wrasse bears little resemblance to the adult, which is greenish brown with scattered blue dots and has a bright yellow tail. Like most wrasses, this species feeds on a variety of small bottom-living invertebrates such as crabs, shrimps and molluscs.

75 Bird Wrasse, male (18 cm)
Gomphosus varius; Labridae
The unusual elongate snout of this fish is useful for searching for food in reef crevices and holes. Prey includes small crabs, shrimps, juvenile fishes, brittle stars and molluscs. Young Bird wrasses are mainly white with a pair of black stripes on the side. Females and young males, known as initial phase fish, are blackish on the rear half of the body and white, overlaid

with black spotting on the front half. The fish shown here is a terminal phase male.

76 Rockmover Wrasse (25 cm)
Novaculichthys taeniourus; Labridae
This fish gets its name from its habit of seizing rocks in its strong jaws and either rolling them over or actually picking them up. This is all part of their hunting strategy. The underside of rocks is home of a variety of food items including mollusc, crabs, sea urchins and brittle stars.

77 Cleaner Wrasse (8 cm)
Labroides dimidiatus; Labridae
Several reef fishes, including members of the wrasse genus *Labroides* feed on the mucus and external parasites of other fishes. Studies have shown these activities are absolutely vital to the overall health of the local fish community. One or more individual *Labroides* occupy permanent territories known as cleaner stations. Members of the nearby fish community, including predators such as groupers, visit them at regular intervals.

78 Bumphead Parrotfish (100 cm)
Bolbometopon muricatum; Scaridae
This is one of the reef's most spectacular inhabitants due to its large size and habit of travelling in graceful schools. Parrotfishes are one of the most important groups of plant-feeding fishes on coral reefs. They play a major role in the energy transfer from plants to animals

73 Humphead or **Giant Wrasse** *Cheilinus undulatus*

76 Rockmover Wrasse *Novaculichthys taeniourus*

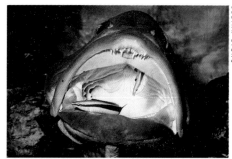

77 Cleaner Wrasse *Labroides dimidiatus*

74 Yellowtail Coris, juvenile *Coris gaimard*

75 Bird Wrasse, male, *Gomphosus varius*

78 Bumphead Parrotfish *Bolbometopon muricatum*

and keep the layer of filamentous algae that covers reefs down to low stubble. The bony hump on the forehead of this species is sometimes used to dislodge large chunks of coral in order to feed on hard-to-reach surfaces.

79 Bluebarred Parrotfish, male (38 cm)
Scarus ghobban; Scaridae

Parrotfishes are closely related to wrasses, but most differ in having the teeth fused into a beak-like structure. Parrotfishes are characterised by female to male sex change. Large males or terminal phase fish are typically much brighter coloured than females.

80 Striped Fangblenny (10 cm)
Meiacanthus grammistes; Blenniidae

The fangblennies are characterised by a pair of enormous canine teeth in the lower jaw which are used for defence. There is a poison gland associated with these teeth and predators quickly learn to avoid the several species of *Meiacanthus* found on the reef.

81 Banded Shrimp Goby (6 cm)
Amblyeleotris sp.; Gobiidae

Although the gobies are the most abundant family on coral reefs, they are less obvious than groups such as parrotfishes, wrasses and damselfishes. However, it is well worth the extra effort to become a "goby watcher". The shrimp gobies have a fascinating relationship with colourful alpheid shrimps, which continually excavate a sandy burrow, providing shelter for both partners. The fish acts like a watch dog, signalling the shrimp with a flick of its tail when the coast is clear.

82 Fire Dartfish (7 cm)
Nemateleotris magnifica; Microdesmidae

Dartfishes are close relatives of gobies. They live in burrows on sand or rubble bottoms. They have a long body and two separate pelvic fins, in contrast to the fused disk-like pelvics found in many gobies. The three species of *Nemateleotris*—of which the Fire dartfish is the most abundant—are among the most elegant of all reef fishes. They are characterised by an elongate, pennant-like first dorsal fin, which is flicked back and forth.

83 Blueband Goby (15 cm)
Valenciennea strigata; Gobiidae

The *Valenciennea* gobies are among the largest and more conspicuous members of the family. Adults are usually seen in pairs on sand and rubble bottoms. They excavate burrows, usually under rock slabs, by removing mouthfuls of sand. During spawning, eggs are deposited in the burrow and guarded by the parents until hatching.

84 Striped Surgeonfish (35 cm)
Acanthurus lineatus; Acanthuridae

Surgeonfishes are named after the scalpel-like blade that folds into a groove on each side of the tail base.

79 Bluebarred Parrotfish, male, *Scarus ghobban*

82 Fire Dartfish *Nemateleotris magnifica*

83 Blueband Goby *Valenciennea strigata*

80 Striped Fangblenny *Meiacanthus grammistes*

81 Banded Shrimp Goby *Amblyeleotris sp.*

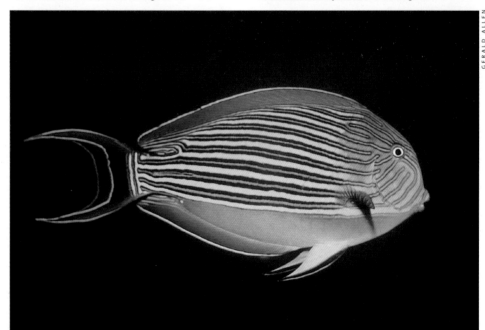

84 Striped Surgeonfish *Acanthurus lineatus*

They can be erected at right angles to the body, presenting a formidable defensive weapon. Surgeons are one of the primary algal feeding fishes on coral reefs often forming huge grazing schools. The Striped surgeon however, is an exception, forming territories that it aggressively guards against all other algal grazers.

85 Spotted Unicornfish (38 cm)
Naso brevirostris; Acanthuridae

The unicorns differ from other surgeonfishes in having one or two fixed spines on the tail base rather than a collapsible spine that folds into a groove. They feed on zooplankton high above the bottom in areas exposed to strong currents. This species has an unusual bony projection above the snout.

86 Palette Surgeonfish (12 cm)
Paracanthurus hepatus; Acanthuridae

This colourful surgeonfish is a shy species, usually seen in small groups on outer reefs or other areas where currents are strong. It feeds in midwater on zooplankton but quickly retreats into branching corals if threatened.

87 Moorish Idol (22 cm)
Zanclus cornutus; Zanclidae

This striking species is one of the first fish that novice divers learn to identify. It is slightly similar to the Longfin bannerfish, but has a longer snout, considerable yellow on the body and a black tail. It is usually seen in pairs or small groups, rarely in large schools. Food items include algae, sponges, sea squirts and other small invertebrates.

88 Orange-lined Triggerfish (22 cm)
Balistapus undulatus; Balistidae

There are 18 different triggerfishes inhabiting reefs of Southeast Asia. They are oval shaped fish with a small mouth, but very powerful teeth. The Orange-lined trigger is usually the most abundant representative found on coral reefs. It feeds on a wide variety of algae and invertebrates including sponge, live coral, crabs, sea urchins, brittle stars, molluscs, worms and juvenile fishes.

89 Foxface or Foxface Rabbitfish (25 cm)
Siganus vulpinus; Siganidae

Sometimes called rabbitfishes or spinefeet, there are 17 species of Signidae in our region, including the Foxface. Live fish should not be handled as all fin spines are venomous. This species is usually seen in pairs. It feeds mainly on sponges and sea squirts but also consumes algae.

90 Clown Triggerfish (12 cm)
Balistoides conspicillum; Balistidae

The attractive Clown triggerfish is usually seen alone, or infrequently in pairs, in the clear water of outer reef slopes. It is a shy fish that quickly retreats when approached too closely.

85 Spotted Unicornfish *Naso brevirostris*

86 Palette Surgeonfish *Paracanthurus hepatus*

87 Moorish Idol *Zanclus cornutus*

90 Clown Triggerfish *Balistoides conspicillum*

88 Orange-lined Triggerfish *Balistapus undulatus*

89 Foxface or Foxface Rabbitfish *Siganus vulpinus*

The **Beaked** or **Longnose filefish** *(Oxymonacanthus longirostis)* is commonly seen among live coral. The species ranges widely in the Indo-Pacific region.

The **Black-spotted pufferfish** *(Arothron nigropunctatus)* is sometimes yellow (page 85) and has a highly variable number of black spots.

91 Titan Triggerfish (50 cm)
Balistoides viridescens; Balistidae

Triggerfishes prepare nests in rocky or sandy depressions. The eggs are guarded by the female until hatching which usually requires less than 24 hours. Female Titan triggers have a particularly nasty disposition and should be given a wide berth when nesting. Its powerful jaws can inflict a nasty bite.

92 Black-spotted Puffer (30 cm)
Arothron nigropunctatus; Tetraodontidae

Puffers or blowfishes are named because of their ability to inflate themselves by swallowing water or air. This habit discourages potential predators. They are also protected by a powerful toxin in their tissues, particularly the liver and ovaries. Under no circumstance should these fish be eaten.

93 Yellow Boxfish (18 cm)
Ostracion cubicus; Ostraciidae

Boxfishes are easily recognised. They have a bony outer covering, no fin spines and a very small mouth set low on the head. They are relatively poor swimmers, having sacrificed speed in favour of protective armour. About eight species occur in our region. The Yellow boxfish has a brightly coloured juvenile stage with a black polka-dot pattern. Older fish are yellowish brown with dark-rimmed white spots. When fully grown the body is purplish brown with very faint spotting.

94 Black-saddled Sharpnose Puffer (8 cm)
Canthigaster valentini; Tetraodontidae

The Sharpnose or *Canthigaster* puffers are miniature blowfishes which have a very limited ability to inflate. About 10 species are known from reefs of Southeast Asia. They are commonly seen in pairs on reefs or surrounding sand and rubble. Their food includes molluscs, crustaceans, worms, brittle stars, bryozoans, sea urchins and sea squirts.

95 Freckled Porcupinefish (22 cm)
Diodon holocanthus; Diodontidae

Porcupinefishes are close relatives of puffers and have the same ability to inflate themselves. They are further equipped with sharp spines. If molested they inflate and the spines protrude at right angles to the body. This is a potent deterrent to most would-be predators, but they are still sometimes found among the stomach contents of Tiger sharks. Five species are seen on coral reefs in our region.

96 Beaked or Longnose Filefish (8 cm)
Oxymonacanthus longirostris; Monacanthidae

Filefishes are very similar to triggerfishes, but usually have a longer first dorsal spine and more laterally compressed body. One of the most commonly encountered coral reef species is the Longnose filefish. It is usually seen in pairs among branched or tabletop corals. It feeds on live coral polyps.

91 Titan Triggerfish *Balistoides viridescens*

92 Black-spotted Puffer *Arothron nigropunctatus*

93 Yellow Boxfish *Ostracion cubicus*

94 Black-saddled Sharpnose Puffer *Canthigaster valentini*

96 Beaked or **Longnose Filefish** *Oxymonacanthus longirostris*

95 Freckled Porcupinefish *Diodon holocanthus*

The **Silvertip shark** *(Carcharhinus albimarginatus)* is potentially one of the reef's most dangerous inhabitants. It is extremely curious and often aggressive towards divers.

The **Whale shark** *(Rhincodon typus)* is the world's largest fish species, but is entirely harmless to humans.

Sharks and Rays
Maligned Veterans of the Deep

Sharks and rays are uniquely specialised fishes that display a primitive stage of evolutionary development. Many of our current genera of sharks and rays originated more than 100 million years ago. Sharks have an extremely ancient lineage and were present in Devonian seas, more than 350 million years ago.

Sharks and rays range from shallow reef flats to deep oceanic trenches many kilometres below the surface. They are common in all seas from polar latitudes to the tropics. Worldwide there are approximately 480 species of rays and 340 species of sharks. Both groups are well represented throughout Southeast Asia. They inhabit a variety of marine environments, and a few also occur in fresh water.

Unlike the vast array of bony fishes, sharks and rays have a skeleton composed entirely of cartilage. Although similar to sharks in many respects, most rays are easily differentiated by the ventral location of the gill slits and by the greatly enlarged pectoral fins, which are fused to the sides of the head. The peculiar sawfishes and guitarfishes have a shark-like body, but the ventral position of the gill openings indicate they are indeed rays. Maximum size is extremely variable. The gargantuan Whale shark reaches a length of 15 metres and the Manta ray may have a wingspan of up to three metres. At the other end of the scale the full grown adults of some deep-sea sharks and electric rays are less than one metre.

Sharks and rays utilise a variety of reproductive modes, but all fertilisation is internal. During the act of copulation the male uses a pair of long, finger-like appendages or claspers to transfer sperm into the female. Depending on the species, the embryos either develop freely, are attached to a placenta, or are sealed in leathery egg cases. Only a small number of species produce egg cases, which are usually deposited in seaweed. Most bear their young alive in broods ranging from a few individuals to nearly 100. The newborn are called pups and have the appearance of miniature adults.

The habits and reputations of a relatively small number of sharks have shaped the popular notion they are evil, menacing predators. This is definitely untrue. The great majority of species are no more threatening than most other fishes. Unfortunately a few species such as the Great white and Tiger shark are known to fatally attack man. Even though the incidence of attacks is small, the danger represented by these animals is consistently blown out of all proportion by the media. Certainly the automobile poses a much greater threat to mankind.

Sharks exhibit a remarkable range of morphological and behavioural diversity. Body shape is often indicative of general behavioural modes. A sleek, well-streamlined shape, such as that of the Silvertip shark (*Carcharhinus albimarginatus*) is characteristic of fast-swimming fish predators. In contrast, the bulky form of the Leopard shark (*Stegostoma fasciatum*) is indicative of a relatively sluggish creature that feeds on less elusive invertebrate prey.

Sharks feed mainly on fishes, crustaceans and molluscs. A few of the larger species consume marine mammals, sea birds, sea turtles and other sharks as well. Fish-eating species generally have well-developed sharp teeth, often with lateral cusps or incisors, that are designed for seizing and tearing. Tooth shapes are very useful for helping to identify individual species. Sharks generally feed at night and have a remarkable adaptation called the *tapetum lucidum* that is also found in cats and other vertebrates that are nocturnal hunters. This structure, located behind the retina, increases the sensitivity of the eye to the available light. Sharks also have an excellent sense of smell and can detect low frequency vibrations at considerable distances helping them locate prey and avoid enemies.

The rays of Southeast Asia are poorly known aside from a relatively small number of reef-dwelling and estuarine species. Most of the common inshore species belong to the family Dasyatidae, popularly known as stingrays. They possess one or more sharp, serrated spines near the base of the tail. These are venomous, causing painful, very dangerous wounds.

Bottom-living rays usually feed on molluscs, crustaceans and other small organisms living in the sand or mud. In order to respire they draw water into the gill chambers via spiracles, a large opening directly behind each eye. This mechanism avoids the ingestion of bottom sediment. Free-swimming rays, such as eagle rays and mantas feed on planktonic organisms and therefore are able to breathe the same way as most fishes—water is ingested through the mouth and passes to the outside through the gill slits.

Below: The sleek Blacktip *(Carcharhinus melanopteros)* is one of the region's most common sharks. It is normally harmless, although attacks, mainly on reef-walkers in shallow waters, have occured.

RUDIE KUITER

**SHARKS
& RAYS**

1 Tawny Nurse Shark (250 cm)
Nebrius ferrugineus; Ginglymostomatidae
This shark shows bottom-resting behaviour that is similar to that of the Reef whitetip shark. However, the Tawny nurse is a bulkier shark and lacks white markings on the fins. It also has a pair of distinctive feelers or barbels, just below the snout. It rests on the bottom during the day and hunts at night. Its crustacean prey is consumed with a powerful sucking action.

2 Silvertip Shark (220 cm)
Carcharhinus albimarginatus; Carcharhinidae
This is an aggressive species that frequently approaches at uncomfortably close range when divers enter its territory. It inhabits outer reef slopes and is particularly prevalent on steep drop-offs, usually below a depth of about 15–20 metres.

3 Scalloped Hammerhead Shark (250 cm)
Sphyraena lewini; Sphyraenidae
Hammerhead sharks have an unmistakable appearance. The hammer-shaped lateral extension on each side of the head probably enhances the manoeuvrability of these graceful swimmers. This modification also increases their powers of vision, smell and pressure detection. Although not normally dangerous, this animal should be respected. Some of the larger hammerhead species are known to attack human beings. The usual diet consists of fishes, crustaceans, turtles and sharks.

4 Grey Reef Shark (180 cm)
Carcharhinus amblyrhunchos; Carcharhinidae
This is the common shark seen on outer reef slopes, in depths of between 5–20 metres. It attains a length of 2.5 m and is considered dangerous, especially when feeding or if unnecessarily provoked. The attack behaviour is stereotypical and involves downward flexing of the pectoral fins and arching of the back. Increased arousal results in exaggerated swinging of the head from side to side. Finally it will repeatedly open and close the jaws and increase swimming speed prior to attack.

5 Blacktip Shark (160 cm)
Carcharhinus melanopterus: Carcharhinidae
This is the most commonly encountered shark in shallow water: they can be found in lagoons, on reef flats, or on coastal fringing reefs. Although it is very inquisitive and often makes close passes, it is generally harmless to scuba divers.

6 Whale Shark (600 cm)
Rhincodon typus; Rhincodontidae
The undisputed heavyweight king of all reef creatures, the harmless Whale shark grows to a maximum size of about 15 m, although animals in excess of 12 m are rarely sighted. Pregnant females may contain as many as 300 young in the uterus. The pups are born alive and measure about 60 cm in length. They grow to nearly 2 m in 3–4 months. Adults are highly migratory

1 Tawny Nurse Shark *Nebrius ferrugineus*

4 Grey Reef Shark *Carcharhinus amblyrhunchos*

5 Blacktip Shark *Carcharhinus melanopterus*

2 Silvertip Shark *Carcharhinus albimarginatus*

3 Scalloped Hammerhead Shark *Sphyraena lewini*

6 Whale Shark *Rhincodon typus*

and their movement is probably influenced by food supply, consisting of small crustaceans, squids and fishes.

7 Leopard Shark (200 cm)
Stegostoma fasciatum; Stegostomatidae

This unusual shark is characterised by a huge tail fin that is nearly equal in length to the rest of the body. It grows to a large size, nearly 350 cm, but is completely harmless. Young sharks under 50 cm are dark brown with yellow zebra-like bars. It is a sluggish animal spending most of the time resting on the bottom.

8 Whitetip Reef Shark (180 cm)
Triaenodon obesus; Hemigaleidae

This inquisitive shark frequently approaches divers at close range but is generally harmless. It spends most of the day at rest on the bottom in caves or under ledges. It is active at night feeding mainly on fishes, but occasional octopus or squid are consumed. Females give birth to litters of between 1–5 pups which are about 50–60 cm long. The maximum adult size is about 210 cm.

9 Coral Catshark (60 cm)
Atelomycterus marmoratus; Scyliorhinidae

This harmless shark is frequently sighted by night divers. It is a small species, attaining a maximum length of only 70 cm. There is very little information about its biology, but it is probably a nocturnal predator of crustaceans and small fishes.

10 Eagle Ray (200 cm)
Aetobatus narinari; Myliobatidae

The Eagle ray is readily recognised by its protruding head, pointed wings, whip-like tail and pattern of white spots on the upper surface. The plate-like teeth are used to crush large shellfish including clams and oysters.

11 Blue-spotted Fantail Stingray (60 cm)
Taeniura lymma; Dasyatidae

The Blue-spotted fantail is one of several stingray species commonly seen in coral reef habitats. The diet consists mainly of molluscs, crushed with its plate-like teeth. Two venomous spines are located on the rear half of the tail. They are capable of delivering a painful wound and therefore this species, as well as other stingrays, should never be handled. It grows to about 70 cm total in length and a disc width of at least 30 cm.

12 Manta Ray (250 cm)
Manta birostris; Mobulidae

Due to its huge size and graceful swimming action, the Manta ray is regarded as one of the reef's prime attractions. They are frequently inquisitive and will linger around a diver for extended periods. The characteristic cephalic lobes or head flaps are used to sweep planktonic organisms, often fishes and crustaceans, into the gaping mouth. Despite their size, mantas are extremely quick and manoeuvrable, and capable of spectacular leaps above the surface.

7 Leopard Shark *Stegostoma fasciatum*

8 Whitetip Reef Shark *Triaenodon obesus*

9 Coral Catshark *Atelomycterus marmoratus*

12 Manta Ray *Manta birostris*

10 Eagle Ray *Aetobatus narinari*

11 Blue-spotted Fantail Stingray *Taeniura lymma*

Mammals
Animals in Need of Conservation

Dolphins or porpoises are frequently encountered in the vicinity of coral reefs. Although seldom seen underwater, their spectacular leaps and bow-riding antics are an integral part of the Southeast Asian diving experience. It's easy to forget these graceful fish-like creatures are mammalian relatives. Although they possess a small amount of hair on the tip of the snout and suckle their young, there are few visible reminders of this relationship. Their sleek, muscular bodies are well adapted for the rigours of marine life. Dolphins can achieve speeds of up to 42 km (25 miles) per hour, exceeding that of most power boats.

Dolphin intelligence has been the focus of considerable research but there is still much to learn. As aquarium visitors know, they can be trained to perform a variety of complex tasks. Although they have relatively poor sense of taste and smell, and limited vision, dolphins compensate for these shortcomings with a sophisticated echo-location system. The sonar not only detects objects at considerable distances but also relays information about their density. The sonar system is particularly efficient as sound travels five times faster in water than on land.

Dolphins live to an age between 20 and 40 years. It takes about five years for a young female to reach sexual maturity and each female is capable of producing a single offspring about every two years. The young dolphin is constantly beside its mother and is suckled for 9–18 months.

The Sea cow or Dugong is another mammal sometimes seen near coral reefs. Although occurring over a wide area between East Africa and the Solomon Islands, it is scarce at most locations. The IUCN (International Union for the Conservation of Nature) lists it as vulnerable to extinction in its *Red Data Book* of endangered species. Fortunately, it is now protected in most places.

1 Spinner Dolphin (200 cm)
Stenella longirostris; Delphinidae
This very sleek and agile species is readily identified by its long narrow snout. It is frequently seen riding bow-waves in all tropical and warm temperate seas. The young are about 80 cm long at birth and adults reach slightly over two metres. It feeds on fish and squid, diving to depths of at least 60 metres.

2 Dugong (150 cm)
Dugong dugon; Dugongidae
The Dugong is a sluggish creature usually found amongst seagrass, which is its main food. While feeding it comes to the surface for air about once each minute but is capable of staying under for up to 6–7 minutes. Dugong may live to an age of 75 years. It takes 10 years for a young female to reach maturity and she gives birth to a single calf about every 3–7 years. Adult dugong grow to a length between 2–3 metres and a weight of 250–400 kg.

3 Bottlenose Dolphin (250 cm)
Tursiops truncatus; Delphinidae
This is the best known species of dolphin due to its ability to adapt readily to captivity and is frequently seen in public aquaria. The species occurs worldwide in tropical and temperate seas. It lives to an age of at least 37 years and grows to about four metres long.

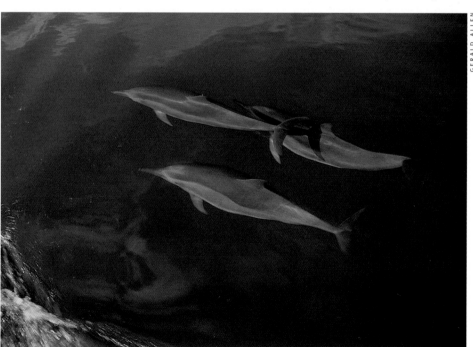

1 Spinner Dolphin *Stenella longirostris*

2 Dugong *Dugong dugon*

3 Bottlenose Dolphin *Tursiops truncatus*

Marine Reptiles

Air-breathing Sea Snakes and Turtles

Although represented by hundreds of species on land and in fresh water, marine reptiles are few and far between, particularly those which are encountered on coral reefs. Turtles are the most conspicuous representatives but several snakes may also be seen.

There are approximately 50 species of sea snakes, all belonging to the family Elapidae. At least two dozen occur in our region, but only the two similar species of banded sea kraits are commonly encountered by divers. They range throughout the tropical Indo-Pacific, but by far the majority of species live close to continental coasts from the Persian Gulf to Southeast Asia and Australia. Tiny Ashmore Reef, lying 150 km south of the Indonesian island of Roti, must surely rate as the world's sea snake capital. Twelve species are recorded from there. A reef walk through shallow water at Ashmore is like negotiating a mine field: well camouflaged and coiled, snakes are everywhere!

Although highly venomous, sea snakes do not normally pose a threat to divers. However, they are often very curious and may approach in a rapid and deliberate fashion that can be easily mistaken for an attack. It's best to remain calm and the snake will eventually continue on its way. On land most sea snakes are relatively helpless and unable to strike. But they definitely should not be handled. The venom is highly potent. The Beaked sea snake, which ranges from the Seychelles to the Coral Sea, is one of the most poisonous. An average sized snake can produce 10–15 mg of venom. A fatal dose for an adult human is only 1.5 mg! Diver fatalities are unknown but snake bite deaths are not unusual in Southeast Asia. The victims are mainly careless fishermen who are bitten while removing entangled snakes from nets.

Sea snakes are air breathers and therefore must periodically return to the surface. There is little accurate information regarding their ability to stay submerged. Snakes have been observed to remain underwater for 10 minutes or so, but experts estimate that some species may remain submerged for several hours. Sea kraits penetrate depths to about 40 metres but, judging from the types of fishes found in their stomachs, it seems that most snakes usually frequent depths between five and 10 metres.

Six of the eight species of marine turtles occur in the Southeast Asian region. Although somewhat diverse in appearance and habits they share certain similarities. All are fully aquatic except for brief periods when the female comes ashore to deposit her eggs. Most turtles are mainly carnivores that feed on a variety of swimming and bottom-living organisms. Such items as jellyfish, sea squirts, sponges, soft corals, crabs, squids and fishes are commonly consumed.

Sea turtles migrate over long distances to reach favoured breeding sites, in some case the same beach or island where they were hatched. Breeding occurs in cycles that vary from about one to five years, but on average a female will breed about once every two years. Nesting frequently takes place during summer or autumn and a single female usually lays several batches of eggs at 2–3 week intervals. A hole is excavated with the hind flippers and about 50–150 eggs are deposited. They are immediately covered with a layer of sand. Incubation time is variable, but for most species lasts about two months.

Hatching usually occurs at night. After emerging from the nest the young turtles must literally run the gauntlet in their dash to the sea. Mortality of the eggs and hatchlings is extremely heavy. Besides man, nest robbers include ghost crabs, dogs, monitor lizards, foxes and monkeys. Those lucky enough to reach the sea face an onslaught of sharks and large fishes from below, and birds of prey from above.

Modern marine turtles are very similar to ancestral forms that shared the seas with ichthyosaurs and pleiosaurs 150 million years ago. Obviously these survivors are extremely well adapted for life in the oceanic realm. Unfortunately, their existence is increasingly threatened. In many areas turtle meat and eggs are utilised for food, the oils are used in the cosmetic industry and for medication, and the shells are shaped into bits of jewellery. As a result humans are their biggest threat. Many turtles are accidentally killed each year by drowning in trawl nets or becoming entangled in set nets. They also fall victim to bits of plastic debris which if mistaken for food can fatally block the breathing and digestive passages. Large, fully grown turtles have few natural enemies. Occasionally they are attacked by sharks and killer whales.

Below: Green sea turtles *(Chelonia mydas)* spend their entire life at sea except when the females come ashore to lay their eggs. Nesting beaches are common throughout the region. One of the best is at Sanggalaki Island, off northeast Kalimantan, where egglaying occurs every night of the year.

BURT JONES/MAURINE SHIMLOCK

MARINE REPTILES

1 Banded Sea Krait (100 cm)
Laticauda colubrina; Elapidae
The Banded sea krait is an active hunter during the day. Stomach contents indicate they feed largely on small gobies that live in burrows. The pattern of banding is very distinctive but some eels, particularly snake eels in the family Ophichthidae, are superficially similar. Sea snakes are easily distinguished from eels, however, by their scales and the flattened, paddle-like tail.

2 Olive or Golden Sea Snake (150 cm)
Aipysurus laevis; Elapidae
This species has the reputation of being very aggressive and dangerous, but reports of attacks are often exaggerated. The species is extremely curious and will rapidly approach a diver from a considerable distance. However, it normally will not bite unless provoked. Therefore, it is not advisable to fend off the snake by kicking or prodding it with a spear.

3 Yellow-bellied Sea Snake (70 cm)
Pelamis platurus; Elapidae
This is the most widely distributed sea snake, ranging from East Africa to the American Pacific. It lives in the open sea, sometimes hundreds of kilometres from the nearest land. Although not normally seen on reefs, it can sometimes be seen from dive boats. The species is easily recognised by its bold colour pattern which is unlike that of any other sea snake.

4 Loggerhead Turtle (80 cm)
Caretta caretta
The reddish-brown colour of the carapace and prominent overlapping beak are useful in identifying the Loggerhead. The Hawksbill turtle has a similar beak but often has overlapping shell plates or scutes. In addition it has four lateral scutes on each side of the shell compared to five in the Loggerhead. The Loggerhead nests on sandy beaches throughout the region.

5 Olive Ridley Turtle (75 cm)
Lepidochelys olivacea
This turtle is similar to the Green sea turtle in appearance but the head is comparatively larger. It has five or more lateral scutes on each side of the central row, and these are much narrower and elongated compared with those of the Green sea turtle. It is estimated that there are some 1,000 nests in the region.

6 Green Sea Turtle (100 cm)
Chelonia mydas
This is the most commonly encountered turtle in the region. It is recognised by its relatively small head and blunt snout that lacks a distinct bill. Adults feed on seagrasses and algae. The Green sea turtle is reported to reach 400 kg, but the largest in Southeast Asia grow to about half this size. Females usually return to the same beach for successive nestings, sometimes actually the same beach where they emerged as hatchlings.

1 Banded Sea Krait *Laticauda colubrina*

2 Olive or **Golden Sea Snake** *Aipysurus laevis*

3 Yellow-bellied Sea Snake *Pelamis platurus*

6 Green Sea Turtle *Chelonia mydas*

5 Olive Ridley Turtle *Lepidochelys olivacea*

4 Loggerhead Turtle *Caretta caretta*

Index